ERRATUM

The description of this book on the inside front flap of the
dust jacket contains an error. The third sentence of the
second paragraph should read:

Khadduri explains how the rise of Persia as a Shi'ite
state helped divide Islam into two factions (Shi'ite and
Sunni Muslims) and how the Sunni-Shi'i controversy inflamed tensions between the two countries.

The publisher wishes to apologize for any inconvenience
resulting from this error.

# THE GULF WAR

# THE GULF WAR

## The Origins and Implications
## of the Iraq-Iran Conflict

Majid Khadduri

New York          Oxford
OXFORD UNIVERSITY PRESS
1988

# Oxford University Press

Oxford   New York   Toronto
Delhi   Bombay   Calcutta   Madras   Karachi
Petaling Jaya   Singapore   Hong Kong   Tokyo
Nairobi   Dar es Salaam   Cape Town
Melbourne   Auckland

and associated companies in
Berlin   Ibadan

Library of Congress Cataloging-in-Publication Data
Khadduri, Majid
  The Gulf war.
  Bibliography: p.
  Includes index.
    1. Iran—Relations—Iraq.   1. Iraq—Relations—
Iran.   3. Iraqi-Iranian Conflict, 1980–
I. Title.
DS274.2.I72K48   1988      955'.054        87-38328

2 4 6 8 9 7 5 3 1

Printed in the United States of America
on acid-free paper

# Preface

The purpose of this work is to discuss the cumulative events and unresolved issues that initially brought about war between Iraq and Iran—two close neighbors linked not only geographically but also by intricate historical traditions—and subsequently led to the involvement of other Gulf countries. I have sought to discuss the major events and issues that have led to almost continuous tension and conflict between the two neighbors ever since Iraq came into existence as a modern state following the First World War. It is, however, not a study in military strategy.

Nor is this volume merely a study in Iraqi-Iranian conflicts. Today, no Gulf country can remain aloof from a major conflict between one member of the Gulf family and another; sooner or later other members will find themselves drawn willingly or unwillingly into that conflict. Moreover, since antiquity domestic Gulf conflicts have tended to invite foreign intervention. For this reason, the last three chapters are devoted to the involvement of other Gulf members in the war; in particular, special attention has been paid to the question of how peace and security would be reestablished in the Gulf. In search of peace and security, not only the Gulf, but also other powers, whether individually or collectively, have become intimately involved through the United Nations. In the last chapter, peaceful endeavors to bring the war to an end are discussed. It is the message of this book, however, as I point out in the final chapter, that though the Gulf War might come to an end at any moment, unless the issues that led to the war are resolved, war might break out again at any moment in the future.

As I did in preparing previous works, during visits to Iraq and other Gulf countries I sought the assistance of several men in high offices who, in our discussions, provided me with information and advice, in addition to the information I obtained from official publications, the press, and studies by scholars and writers of various shades of opinions. Although my visits to Iran preceded the revolution of 1979 (my last visit was in June 1977), I have consulted several Iranians who left their country after the revolution in some of the Arab and Western capitals. Needless to say, I have been under no illusion that the men whom I interviewed or consulted have not often given me their own personal views and opinions about events and issues. Not all those whom I had the privilege of interviewing or consulting have permitted me to identify them as sources of information, but the names of those who have given me permission may be found in the footnotes. I should like to acknowledge the kindnesses extended to me by all from whom I had the privilege of seeking assistance. Finally, I wish to thank the School of Advanced International Studies of Johns Hopkins University for extending to me secretarial assistance, and my daughter, Shirin Ghareeb, for compiling the index. No one, however, is responsible for any error or for personal opinions expressed in this work.

M. K.

# Contents

# A Note on the System of Transliteration

Western writers have used different systems of transliteration in reproducing names and words from non-Western languages. As in my earlier works, I have followed the system adopted by the Library of Congress and the editors of the Encyclopaedia of Islam with slight variations, particularly in omitting the diacritical marks and letters at the end of words that are not pronounced in the original language, such as Makka, Madina, and Basra (instead of Makkah, Madinah, and Basrah). But commonly known place names that have been anglicized, such as Persia (for ancient Iran), Syria (for Suriya), Algeria (for al-Jaza'ir), and Cairo (for Qahira), have not been transliterated. However, I have adhered to the phonetic system for the spelling of such names as Khumayni (Khomeini), Husayn (Hussein) and Asad (Assad).

# THE GULF WAR

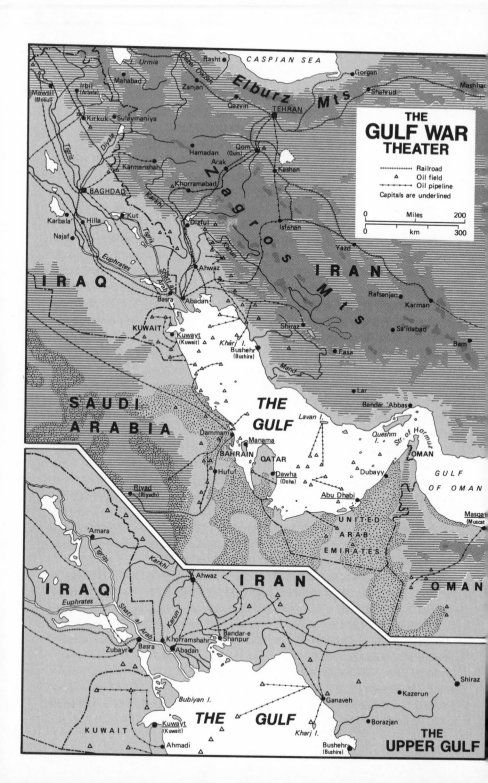

# I

# Introduction

The Gulf War, which began as a conflict over territorial sovereignty and frontier security between two Gulf countries, has now begun to engulf almost all of the Gulf countries to varying degrees. Why has this war, it may be asked, proved so difficult to control and bring to an end, if it is being fought merely over territorial matters? Writers and diplomats, viewing the war from various perspectives, have given us differing interpretations of the origins and drives of the war. Is it really a conflict over frontiers and territorial sovereignty, as it has been dealt with by diplomats at international councils, or is it the projection of an urge for domination over a region potentially rich in oil and strategically vital in any possible conflagration in which the major powers of the world might be involved? Are the confessional (sectarian) divisions in Islam, to which the majority of the people of the region belong, the root cause of the conflict, or are they merely the rationalization of deeper historical events and traditions which consciously or unconsciously prompted rival rulers to engage in competition and conflict? How much should the foreign powers, let alone the superpowers, get involved in this conflict, and has their involvement aggravated or reduced the dimensions of conflict? Did international organizations—global or regional—which often exercise certain functions of a world government ever attempt to resolve some of the issues or contribute to bringing them under control?

Before we address ourselves to the major Gulf issues, perhaps a few words about the historical and ethnocultural identities of Iraq and Iran might be useful to bear in mind, as the identity of each had a significant influence on the role played by each when they passed under Islamic rule. Iraq developed a civilization (the Babylonian-Assyrian) and a highly organized form of government long before Persia (Iran's historic name) had come into existence. But by the seventh century A.D., while Persia struggled to preserve its essential character after becoming part of the newly established Islamic Empire, Iraq's identity had almost completely vanished, and its people were assimilated by the culture and had adopted the language of Arab rulers. Iraq regained its place of prominence when Baghdad became the seat of government and the center of Islamic learning in the middle of the eighth century A.D. After the fall of Baghdad in 1258, Iraq was subjected to migrations and foreign domination—Mongol, Saljuk, and Ottoman—and remained under Ottoman dominion for over four centuries. Not until the end of the First World War did Iraq emerge as a new nation-state.

Persia's separation from the Islamic Empire began much earlier than Iraq's, when, in the early sixteenth century, it reappeared on the political map of Islam as an independent state and adopted the Shiʿi heterodox creed as an official religion. It stood in opposition to the Ottoman sultans, who became the spokesmen of the Orthodox Sunni creed, the major division of Islam. The warfare that ensued between the two Islamic states, outwardly over confessional issues, was later transformed into conquest and rivalry between sultans and shahs over domination and control of Islamic lands. Iraq, where the original schism in Islam began, inherited both the historical confessional controversy and the subsequent conflict over territorial sovereignty and security as a successor to the Ottoman Empire.

The people of Iraq, at least when the country came into existence after the First World War, were split almost equally into the two Sunni and Shiʿi communities. The Shiʿi community, which had been discriminated against under Ottoman rule, began to play an increasingly important role under the newly created political system after the country's separation from the Ottoman Empire. Owing to the deep confessional loyalty of Shiʿi followers, relations between Iraq and Persia were necessarily affected by the domestic tensions between the Sunni and Shiʿi communities. This situation has become even more crucial in the relationship between Iraq and Iran since the Iranian Revolution and the establishment of the Islamic Republic, claiming to assert the "true Islam" (Khumaynism), in 1979.

This book is neither an essay on military history nor on the conduct of the on-going military operations. Rather its aim is to inquire into the causes of the war, first as a conflict between two, and later other, Gulf countries, and to explore how it might be contained and brought to an end by peaceful means.

Since some of the causes that brought about the conflict originated long before the two major parties became sovereign states, an inquiry into the schism in Islam that created the estrangement between the Sunni and Shiʿi communities is necessary before the more recent territorial and security problems can be tackled. The confessional or communal conflict has permeated Perso-Iraqi relations to the present day, even though both countries have adopted nationalism, presumably to supersede confessional loyalties, as a symbol of identity in both bilateral and international relations.

It is proposed in this study to pursue more than one method. Some thinkers, who assume that the drama of history is the unfolding of an already prescribed set of principles and rules of human conduct, whether of divine or human origin, maintain that the movements and events we are witnessing today should be viewed and interpreted in accordance with the standard of their own religious-political system. If the outcome of events does not conform to their perceived standard, it is held, the whole process is condemned to have fallen under evil influences—Satanic or others—and therefore should be stopped or corrected. The Ayat-Allah Khumayni and his followers—indeed, all other believers, often called Muslim "fundamentalists"—fall into this category.

There are, on the other hand, some modern writers inspired by the economic interpretation of social and political movements, who have tended to interpret social change on the grounds of shifting social and economic conditions. Indeed, some have gone so far as to interpret all social movements in the Middle East in accordance with social and economic doctrines. By providing statistical data about conditions, these writers have ascribed all revolutionary changes to socioeconomic causes, even though it is well-known to experts in the field that statistical data are incomplete and often faulty and unreliable. These social analysts, considered by their Western peers to have offered plausible arguments, have in fact provided inadequate and often misleading interpretations. Scholars who know Islamic society better have long realized that believers did not always respond to the call of leaders in accordance with socioeconomic doctrines but to other more intricate and deep-rooted traditions. Believers in Islamic lands are still duty-bound to respond to the call of the ʿulama, especially Shiʿi mujtahids, to whom they owe their loyalty in principle, irrespective of shifting social and economic conditions. The Imam, to Shiʿi followers, is infallible and his authority is ultimately derived from God.

It is proposed in this study to combine the advantages of the so-called idealist approach without disregarding the advantages of the realist method. An attempt to combine idealism with realism, which may be called "empirical idealism" is a method I have pursued in earlier studies. It is thus not our purpose to elucidate only norms and values in abstract terms, but to relate them to movements and men who strive to realize them under existing conditions.

# II

# Persia's Rise as a Shi'i State and Its Conflict with the Ottoman Empire

Iraq and Iran, forming the eastern front of the Islamic Empire, were both occupied at an early stage of Islam's expansion beyond Arabia during the mid-seventh century, and both made no mean contributions to the development of Islamic civilization. Iraq, however, was thoroughly assimilated by and identified with the early Arab conquerors who established the Islamic Empire. It eventually became an integral part of the empire, partly because its ancient civilizations had vanished and partly because its southern area was inhabited by Arab settlers who had migrated there—long before Islam arose—in large numbers and were ruled by Arab princes. Persia, on the other hand, different in ethnocultural background, resisted the wave of Islamic expansion and clung to its language and historical traditions, though the great majority of its people were converted to the Islamic faith.

Since Persians reluctantly submitted to Islamic rule and resisted assimilation, they were often accused by Arabs of supporting heterodox movements and of consciously introducing unorthodox elements into the creed hoping to alter the structure of Islamic society. Arab writers in particular reproached their peers of cultural bias, often denounced as ethnicists (shu'ubiyun), because they sought to glorify Persian history and culture at the expense of the Arab heritage. Above all, they were accused of espousing Shi'ism (a heterodox division of Islam), into which they had injected Persian notions and practices either to undermine the Sunni creed (to which the majority belonged) or to claim Shi'ism as their own religion.

But Shi'ism was neither founded by Persians nor did it spread widely into their country until the sixteenth century, when it was finally adopted as the official religion of the state by Shah Isma'il, founder of the Safawi dynasty, and became instrumental in the separation of the country from Islamic unity.

6

Earlier Shi'ism was a protest movement that had arisen not in Persia but in southern Iraq, whose center was Kufa, a town on the lower Euphrates, to which 'Ali (d. 661), the fourth Caliph, had moved his seat of government to rally support from its Arab settlers in his struggle for power against Mu'awiya, Governor of Syria, who represented the tribal aristocracy of Arabia.[1] Arab supporters of the fourth Caliph were called the " 'Alids," but later came to be known as the Shi'a (partisans) of 'Ali. After 'Ali's assassination, Mu'awiya assumed the Caliphate, and 'Ali's two sons—Hasan and Husayn—retired to Madina, Islam's first capital, until the death of Mu'awiya in 680.

Even before the death of Mu'awiya, the cause of Shi'ism had gathered momentum because of the repression and persecution of Arabs—suspected of 'Alid sympathy—in southern Iraq at the hands of brutal governors appointed by the Caliph from Damascus. After Mu'awiya's death and the assumption of power by his son, Yazid, 'Ali's second son, Husayn (Hasan, the older brother, had died in 669) began to claim his right to the Caliphate. The 'Alids, his father's followers, sent messages to Husayn, urging him to come to Kufa, his father's seat of government, hoping that he would lead the opposition against Yazid, who had been proclaimed Caliph. As he approached Kufa, Husayn was intercepted on the outskirts of Karbala (near the River Euphrates) by a small force sent by the Governor of Iraq to prevent him from proceeding to Kufa. Upon refusing to return to Madina, as the governor's men commanded, he fell in his encounter with the intercepting force. His death at Karbala on October 10, 680, commemorated every year by Shi'i followers on the tenth of the month of Muharram, is an occasion of mourning and repentance for the fall of Husayn. The tragedy of Karbala is considered the *cause célèbre* of justice to the Shi'a all over Islamic lands in their struggle to achieve power under the banner of a Shi'i Imam (ruler) against their opponents (later known as Sunni followers), who claimed that the caliph should be nominated and enthroned not by legitimate right but by the consent of the people.

Under Umayyad rule an increasing number of Persian converts, discriminated against in taxes and service in the state, began to support the Shi'i movement, and their support contributed in no small measure to the downfall of the Umayyad dynasty in 750. The opponents to Umayyad rule consisted, in the main, of a coalition of the descendants of 'Ali and the descendants of Ibn 'Abbas ('Ali's uncle), who cooperated in the destruction of the Umayyad dynasty, presumably so that power would be entrusted to a Shi'i candidate. When, however, the last of the Umayyad caliphs fell, power was seized by one of the descendants of Ibn 'Abbas, called al-Saffah, who became the founder of a new dynasty, the 'Abbasid, and Shi'i followers were again deprived of the fruits of their endeavors. Persian supporters were amply rewarded, not only by the office of First Minister (Wazir, or Vizir) but also by other high offices which were earlier denied to them.

Shi'i followers, however, never giving up their support to the claim of Husayn's descendants to the Imamate, continued to challenge Sunni rule. In time their movement was augmented by an increasing number of Persians, especially as the 'Abbasid dynasty began to decline and the Shi'i cause gained popularity and prestige. Their propaganda was so effective that they succeeded in establishing three Shi'i dynasties before the fall of the 'Abbasid regime to the Mongol invaders in 1258. One was in Egypt, called the Fatimid dynasty, which lasted from the tenth to the twelfth century; the second, the Hamdanid, ruled over northern Syria and Iraq in the tenth century; and the third, the Buwayhid, ruled over southern Iraq and West Persia (Daylam) in the tenth century. The Buwayhid dynasty seems to have preferred to rule over a majority of Sunni followers in the name of Sunni caliphs, rather than to impose Shi'i rule, as their contemporary Fatimid Imams had in Egypt. Having ruled for two hundred years, the Fatimid dynasty was overthrown by Salah al-Din (Saladin), who re-established Sunni rule in Egypt in the latter part of the twelfth century.

The increasing infiltration into Persia and Asia Minor of Saljuk and Turkish tribes (recent converts to Sunni Islam) from Central Asia—not to speak of Sunni Mamluk rule in Egypt and Syria—contributed in no small measure to the consolidation of Sunnism. It was, however, the establishment of the Ottoman dynasty in Asia Minor at the opening of the fourteenth century, and its adoption of the Hanafi School of Law (one of the leading Sunni schools) as the official school of the state, which consolidated Sunnism in areas where Shi'ism had become firmly established. No less important was the spread of Sunni Islam westward by Ottoman conquests, at the expense of the Byzantine Empire and of Christian princes in Central Europe, which aroused the Sunni circles admiration for the achievements of the Ottoman sultans, just as they had taken pride in the early caliphs who founded the Islamic Empire in the seventh century. Small wonder that when the Ottoman sultans turned eastward to extend their control over the Arab provinces, they were welcomed by the majority of the people—though not by their Mamluk rulers—who did not consider the Ottoman sultans foreign conquerors, but rather hailed them as their legitimate rulers and the bearers of the banner of Islam against infidels.

Toward the end of the fifteenth century a resurgence of Shi'ism in Persia had a far-reaching effect on Shi'i followers in other lands, since the establishment of erstwhile Shi'i dynasties in Arab lands proved to be of relatively short duration. Indeed, Shi'i followers had never given up their claim to the legitimate Imamate, even after their Twelfth Imam had suddenly disappeared in 874, when he was still an infant. In their eyes his absence (ghayba) was only in body, since his spirit was considered still with the fold, and it is believed he will eventually return in the capacity of Mahdi (Messiah) to re-establish Shi'i ascendancy and justice. This Messianic notion fostered, in Shi'i circles, the expectation that eventually the time would come when their position would be enhanced and justice reestablished.

The emergence of Persia as a Shi'i state was thus hailed as a sign that the star of Shi'ism had again risen and that power, at last, had been restored to them. Persia was perhaps the most suitable country for such a resurgence; for, though Shi'i followers were still a minority, the people as a whole took pride in the historical memory that their country was once a great empire, and the Persian language, though diluted with innumerable Arabic words and written in Arabic script, continued to serve as the national language and had even become the *lingua franca* in northern and eastern Islamic lands. What set all of these forces in motion was the founder of a new Persian dynasty, who seized power in the name of the Imam and established Persia as a separate state on the political map of Islam, with Shi'ism as its official religion. The rise of Persia became a challenge first to Sunnism under Ottoman rule and later to Sunni rulers in Iraq and beyond in other Arab lands.

The founder of the new Persian dynasty was Shah Isma'il al-Safawi. Born in 1486 to a family that combined the traditions of mysticism and militarism, Shah Isma'il exploited at the early age of thirteen the fanaticism of a group of Qizil Bashi (redheads) which his father had organized at Gilan in northern Persia to spread Shi'i teachings.[2] Possessing prudent judgement and the qualities of a charismatic leader, he exploited the rivalry among Sunni leaders and was able not only to establish control over the whole of Persia, but also to extend it into Central Asia. Above all, he sought to impose Shi'ism upon his countrymen by force and to extend his rule in the name of the Imam to other Islamic lands. His influence and prestige rose so high that rival Turkish Sunni rulers in Central Asia sought his support by offering to favor Shi'i followers in their provinces. In his drive to reassert Shi'ism and spread it beyond Persia, he instructed his followers to curse not only the Caliphs Abu Bakr and 'Umar but all Sunni caliphs, and to venerate Husayn, who died in his struggle to regain the Imamate from Sunni rule in 680, with an annual funeral procession and other ceremonies (al-mawakib) on the tenth of the month of Muharram. Several other rituals which Shi'i followers perform today, such as the annual visitation to the sanctuaries of Najaf and Karbala and the reassertion of the jihad, have become part of Shi'i teachings and traditions. Moreover, the scattered Shi'i communities in predominately Sunni provinces often looked to Persian authorities for support in their struggle against Sunni domination.[3]

Apart from his prudence and leadership, Shah Isma'il was initially successful in establishing control over Persia and Central Asia because of the absence of a powerful rival from the political stage. The policy of the Ottoman Empire had been not to extend its rule eastward into Islamic lands, but to penetrate westward into Europe at the expense of Christian princes. When the Shi'i communities in Asia Minor and other Ottoman provinces became restless, however, (under the influence of Shi'i instigations from Persia) and showed signs of rebelling against Sunni domination, Sultan Salim, who had just seized power from his father in 1514, turned eastward not only to suppress Shi'i rebellions in his own dominions, but also to reestablish Sunnism in the areas of Islamic lands where Shi'ism had prevailed. Like Shah Isma'il, his ambition was not merely to be sultan over the lands under his control, but the supreme head of the whole Islamic world.

Renowned for his valor and ruthlessness, Sultan Salim crushed the Shi'i rebellion in eastern Asia Minor and put to death almost the entire Shi'i community without mercy. From there, he proceeded eastward and captured Tabriz, capital of Shah Isma'il, and defeated his forces at the battle of Chaldiran (1514). But, perhaps to the surprise of his opponents, he did not pursue the Shah's forces, which retreated to the interior of Persia, and instead turned seeking glory to the Arab countries. Almost without resistance—indeed, he was welcomed by the people as the champion of Sunnism against Shi'ism—Sultan Salim occupied the whole of al-Jazira (northern Iraq), Syria, and Egypt in 1515–17, leaving Baghdad, Najaf, and Karbala, which Shah Isma'il had occupied in 1508, under Persian control. Had he pursued Shah Isma'il before turning to Arab lands, he would probably have been able to restore Persia to Sunni rule and reestablish Islamic unity; but his change in plan, presumably because of the rugged country and harsh winter, gave Shah Isma'il an opportunity to regroup and enlarge his military forces and to recover not only Tabriz but the whole area that had been earlier under his control.

Before his confrontation with Sultan Salim, Shah Isma'il seems to have envisioned the re-establishment of Islamic unity under Shi'i control. But with Sultan Salim's failure to pursue his victory over Shah Isma'il, the house of Islam was virtually divided into two communities based on territorial segregation. A vast central strip between the two territories consisting mainly of central and southern Iraq, became the bone of contention between Persian shahs and Ottoman sultans. The Shi'i sanctuaries of Najaf, Karbala, Samarra', and Kazimayn, where Shi'i followers had long been residing, were in the Middle Euphrates area. Yet the Ottoman sultan seems to have been determined to keep all Arab lands, including the Middle Euphrates area, under his control, since the majority of the population of Southern Iraq were Sunni followers.[4] Before he could recover the Arab provinces which Shah Isma'il had occupied, Sultan Salim died suddenly, shortly after his return from Egypt in 1520. He was succeeded by his son Sultan Sulayman, known to Europe as the Magnificent. Shah Isma'il died four years later, in 1524, without realizing his dream of extending his control over the entire Gulf region.

Although the circumstances for recovering Baghdad and the Middle Euphrates area from Persian control were not unfavorable, Sultan Sulayman's preoccupation with his European campaigns kept him from pursuing his father's policy of extending Ottoman control eastward for almost a decade. It was in response to Sunni appeals, after Abu Hanifa's mosque and other sanctuaries in northern Iraq had been pillaged by Shi'i followers, that Sultan Sulayman was finally stirred and turned his operations eastward. In brilliant campaigns, beginning with Tabriz, he marched southward toward Baghdad, and quickly recovered the whole area from Azirbayjan to the Middle Euphrates from Persian domination in 1534. For over two centuries, the Shi'i sanctuaries in Najaf, Karbala, Kazimayn, and Samarra' (to which Shi'i followers perform visitation) were in the Sunni's hands, under whom the Shi'i community suffered repression and discrimination. An account of Persian attempts to recover control over Baghdad, Karbala, and Najaf belongs to military history. Suffice it to say that, despite several attempts by Shah 'Abbas the Great and Nadir Shah, though Baghdad and the Middle Euphrates areas were occupied by Persian forces (the former in 1623 and the latter in 1743–44), Persian occupation proved temporary. In time Ottoman control over the three provinces of Mawsil, Baghdad (in which the Middle Euphrates area was included), and Basra was firmly established. Several treaties, intended to establish peace and delimit frontiers, were concluded, but to these we shall return.[5]

The foregoing account of the Perso-Ottoman rivalry is intended to provide the reader with the historical background of the division in the house of Islam, inherited by both Iraq and Persia, which may throw light on the events that are now taking place in the Gulf region and beyond. Its significance may be summarized as follows.

First, the territorial division of Islam on confessional grounds was not new, as noted earlier in the case of Egypt under Fatimid rule; but the ascendancy of Persia as a Shi'i state, accompanied by territorial segregation, has become firmly established since the beginning of the sixteenth century and has continued to divide the Islamic body politic. The territorial segregation of Islam was accomplished through terrible bloodshed; yet it was necessary if the emerging countries were to survive and later be integrated as independent states in the larger world community of nations. It coincided with the breakdown of Western Christendom, a prototype of ecumenical society, which was transformed first into a European state-system and later into a world community of nations. But the division of Islamdom on sectarian grounds, though providing the rationale for the separation of Persia from Islamic unity, did not set a precedent for the rise of other Muslim states on confessional grounds. After the First World War, when Persia and Iraq became members of the modern community of nations, they had already adopted nationality as their symbol of identity.

Second, the division of Islam into several political entities in the sixteenth century did not necessarily mean that peace had prevailed among them. It is true that the Ottoman and Persian states sought at the outset only to end hostilities, but not necessarily to make peace, until they had finally agreed to establish permanent frontiers and put an end to differences over territorial matters, as clearly envisaged in the elaborate peace treaty of Arzurum (Ard al-Rum) in 1847. This treaty, the first concluded between two Muslim sovereign states in accordance with Western rules and practices, was not the first to establish peace on the grounds of territorial segregation; it was preceded by several others that were concluded in accordance with Islamic law, beginning with the Treaty of Zuhab (1639). The Treaty of Arzurum, concluded to establish peace and delimit frontiers, became the foundation and principal legal precedent for all subsequent dealings between the two Muslim sovereign states. Meanwhile, the two rival Muslim states had been drawn into the European system of the balance of power, and their increasing commercial and diplomatic relations led to their final admission into the modern community of nations. Even before the new state of Iraq had come into existence after the First World War, Persia and the Ottoman Empire had already adopted modern principles governing the relations among states. Iraq, however, a successor state, inherited both the legal and historical legacy of the Ottoman Empire in its relations with Persia.

Third, since the Ottoman and Persian Empires agreed to make peace and recognize the sovereignty and territorial integrity of one another, it was almost taken for granted that they would take formal steps to set aside their creedal differences in the conduct of foreign affairs—not to speak of domestic affairs—just as the Christian states of Europe had agreed to relegate religious differences to the domestic level following the Treaty of Westphalia (1648).[6] Indeed, Nadir Shah, in a treaty with the Ottoman sultan (1763), proposed a compromise formula which would transform Shiʿism into a fifth school of law called the Jaʿfari school, but the sultan did not ratify the treaty on the ground that it was opposed by Ottoman Sunni scholars, although Sunni and Shiʿi scholars in Persia and Iraq seem to have been agreeable to it in principle.[7] In time, the Ottoman and Persian states only tacitly agreed to relegate religious differences to the domestic level, as they gradually came to participate in international councils and enter into treaty arrangements with European states, although Sunni and Shiʿi followers were never really reconciled to the idea of a national identity in their inter-Islamic relations.

Fourth, both the Ottoman and Persian Empires agreed to establish permanent diplomatic missions in each other's capital—another sign that they accepted the condition of peace to prevail between them—since both had already agreed to exchange permanent diplomatic missions with European powers and to conduct their foreign relations in accordance with the practices of European diplomacy.[8] Moreover, since the Ottoman and Persian Empires had entered into alliances with several European powers, especially Britain and Russia, and accepted European good offices to resolve their frontier disputes, it was taken for granted that Western legal principles and practices applicable to such disputes would be acceptable to them.

Finally, upon their admission into the modern community of nations, the Ottoman and Persian states became in time subjects to International Law, perhaps as early as the mid-nineteenth century. Before they became members of the community of nations, however, they were considered outside the pale of the law of nations. With Persia's rise as a rival to the Ottoman Empire in the early sixteenth century, Spain, England and other European powers extended diplomatic support to it, just as the king of France's rivalry with other Christian princes had induced him to enter into an alliance with the Ottoman sultan and make peace with the most powerful Muslim state in 1535.

The two Muslim states were thus drawn into the European balance of power by rivalry stemming from creedal differences, and their increasing commercial and diplomatic contacts with European countries led to their final admission into the modern community of nations. Thus, when Iraq came into being after World War I, Persia and the Ottoman Empire had already agreed to govern their bilateral relations as well as their relations with other states of the world in accordance with modern (secular) principles and practices. Iraq, as a new state that emerged from the dissolution of the Ottoman Empire, inherited not only the assets but also the liabilities of the mother country.

# III

# The Iraq-Iran Conflict During the Inter-War Years: The Confessional Aspect

Following the First World War, the dissolution of the Ottoman Empire, whose fall had long been expected, changed significantly the political map of the Islamic world. Not only Turkey, the mother country, but also other countries were created as new nation-states on the basis of national rather than ecumenical (Ottoman) identity, which had recognized the individual as a citizen according to his religious affiliation. In accordance with the Treaty of Lausanne (1924), Turkey renounced its claims to territories that were inhabited by non-Turkish population, resulting in the detachment of all Arab countries (the lands inhabited by a majority of Arabic-speaking peoples) which either emerged as independent states, such as Najd (later Saudi Arabia) and Yaman, or passed under temporary foreign control, like those in northern Arab lands, before they became independent. Turkey, remaining in possession of a relatively small portion of Ottoman territory, was able to put an end to its historic conflict with Persia, since neither the new frontiers that separated the two countries were controversial, nor did any significant portion of the Shiʿi community remain within its territory.

Iraq, one of the Arab successor states, is the only country that inherited the rest of the frontier between Persia and the Ottoman Empire. Not only did it inherit that portion which had not yet been fully demarcated, but it also inherited the former provinces of Mawsil, Baghdad, and Basra, which had become the battleground of almost continuous military operations since the sixteenth century, before the two countries agreed to put an end to hostilities and settle their differences peacefully. Perhaps no less important, the Shiʿi community, which had been under Ottoman rule as part of the Baghdad province, became an integral part of the people of Iraq, and the Shiʿi sanctuaries in Najaf, Karbala, Kazimayn, and Samarra', to which the Shiʿa

14

from Persia and other Muslim lands perform periodic visitations, passed under Iraqi control. Since the Shi'i community was a minority under Ottoman rule, it suffered discrimination and repression and could not possibly enjoy an equal status with their Sunni Arab compatriots, although the Sunni Arabs of Iraq were not in practice granted all the privileges enjoyed by their Turkish co-religionists. But Sunni Arabs had access to public education and service in the army and the civil administration, which the Shi'a were denied. For this reason, the Shi'a were bound to fall back on private religious schools to educate their children, and most of the well-to-do families were engaged in business. The rank and file were illiterate, and the majority of the Shi'a in the Middle Euphrates area were still leading a nomadic life in the desert. Some of the Shi'a in the urban centers of Iraq were of Persian descent (most of them resided in Karbala and Kazimayn),[1] but the majority were of Arab descent, although intermarriage was common and trade relations between Persia and Iraq were almost the monopoly of Shi'i businessmen. The Shi'i community made up about half of the population of the country, whose numerical strength was less than three million when it was separated from Turkey. Small wonder that when Iraq passed under British control the Shi'i community felt relief from Ottoman domination, particularly when Britain made a number of public statements, vague as they may have been, promising self-government and independence. Since the Shi'a formed the largest single confessional community in the country, it naturally sought to improve its condition, and those who considered the Shi'a to form the majority of the population expected to play a more meaningful role in governing the country, if not the predominant role.[2]

British policy, however, ran contrary to Shi'i expectations. When Britain occupied southern Iraq shortly after the outbreak of the First World War, no clear policy had yet been formulated about the future of the country. Such a policy began to evolve as the country gradually passed under British control by the end of the war. There were indeed two schools of thought concerning the future of Iraq and the Gulf region. The first, often called the Colonial Office school, argued that since the British expeditionary force which occupied Iraq was dispatched and controlled from India, Iraq's future should be determined largely by the Colonial Office (as part of the Gulf and the Indian Ocean). The second, called the Foreign Office school, maintained that the increasing British influence in the Eastern Mediterrean and the Red Sea required the support of leaders in areas where Arab nationalism had already begun to develop; it saw in the Sharif Husayn of Makka and his sons the potential rulers who would provide leadership for the Arab nationalist movement as the new ally of Britain. Iraq was considered the Arab country most affected by the Arab nationalist movement and was, indeed, included in the general promise of independence given to the Sharif Husayn when he entered the war as an ally of Britain in 1916. Moreover, some of the Iraqi young men who received their education in the Ottoman capital had already entered the service under the leadership of Sharif Husayn, hoping that their country would become independent.

Before the leadership of Iraq was entrusted to Sharifian hands, however, the country was briefly under British control, which was administered in accordance with the viewpoint of the Colonial Office. The Shi'i community in the Middle Euphrates at first responded favorably to the British administration, hoping to control (at least in its area) the administration of the country after the British military withdrawal. Likewise, Sunni leaders in Baghdad became restless soon after the war came to an end. But Britain had not yet made up its mind about the future of the country, despite the indefinite statements about self-government and independence. Since Arab nationalists in Syria, under the leadership of Faysal—son of the Sharif of Makka—were demanding independence and agitating against French control, some of the Sunni and Shi'i leaders in Baghdad also began to agitate against the British administration and demanded independence. The Shi'i tribal leaders in the Middle Euphrates, at the instigation of the mujtahids in Najaf, Karbala, and Kazimayn, such as Shaykh Taqi al-Shirazi, Abu al-Hasan al-Isfahani, and Shaykh Mahdi al-Khalisi, denounced the British administration. The unrest which had begun in the tribal areas spread throughout the middle and southern parts of the country, culminating in the outbreak of a revolt in June 1920, which lasted over five months. Reinforcements were rushed from Persia, and the British authorities suppressed the revolt by force. Meanwhile, the mujtahids in Persia were also inciting the public to rebel against British control of their country and, like their counterparts in Iraq, sought to entrust leadership to clerical hands.

In 1921, upon the recommendations of a conference held in Cairo, Britain decided to establish a national regime in Iraq, and the throne of the country was offered to the Amir Faysal, son of King Husayn, the Sharif of Makka.[3] Although the revolt in the Middle Euphrates was suppressed, the Shi'i leaders maintained that Britain had decided to abandon its policy of direct control in favor of an Arab government mainly because of the Shi'i resistance to British rule and participation in the nationalist activities of the country. They were, however, in favor of the candidacy of Faysal for the throne of Iraq and hoped that they would play a significant role in his regime.[4]

The choice of Faysal as head of state was perhaps the best possible compromise in principle if a balance between the traditional Sunni leadership and Shi'i aspirations to participate in the country's governance were to be maintained. Since Faysal was the great grandson of 'Ali, the first Imam, according to the Shi'i doctrine, he was agreeable to the Shi'i community as head of state, even though his family had accepted the Sunni creed. To the Sunni community, Faysal was not only agreeable on the grounds of his Sunni affiliation but also because of the leadership which he and his father had provided for the Arab nationalist movement. After Faysal's ascendance to the throne, however, the Shi'i community seems to have had mixed feelings about Faysal's attitude, as he vacillated between British policy, supported by moderate Sunni leaders, and Shi'i pressures to follow an independent policy. Not only was the head of the government, Sayyid 'Abd al-Rahman al-Naqib, a Sunni dignitary, but the Shi'i ministers in the Cabinet were confined to one portfolio—the Ministry of Education. The Shi'i leaders, who congratulated themselves for having contributed to Faysal's elevation to the country's throne, came to the conclusion that the King must have succumbed to British pressures. Since the country was still under British control, Britain was held responsible for Shi'i subordination to Sunni domination, and the Shi'i leaders continued to attack the British administration of the country. Although a few Shi'i leaders, Khalisi and other extremists, were disenchanted with King Faysal, others did not give up hope that he might be persuaded to accept their demands.

The events in Iraq were not unaffected by political developments in Persia. Lord Curzon, British Foreign Secretary, envisaged the establishment of a "chain of buffer states" (Turkey, Persia, Afghanistan, Iraq, and others around the Gulf) to reduce Russian influence in the region and protect India and the Gulf countries. Since Curzon's policy failed in Persia, British policy toward Iraq was reshaped by Winston Churchill, Colonial Secretary, who succeeded in establishing a nationalist regime (often called in England an "Arab façade") which virtually insured the survival of British influence for another decade or two.[5] The Shah of Persia, however, supported by Shi'i mujtahids, took a critical attitude toward British policy and refused to recognize the Iraqi regime until it achieved independence. The Shi'i leaders in Iraq sought to enhance their position by the support they received from Persia.

Following the establishment of a national regime in Iraq, both Iraq and Persia sought through negotiations to resolve the problems they had inherited from the past. The major problems were four: The Sunni-Shi'i tension; Iraq's recognition by Persia; the dispute over frontiers; and the Kurdish problem. Since none of these problems has yet been fully resolved to the satisfaction of both sides, I propose to discuss the root-causes of the differences and the steps taken to resolve them. In this chapter the Sunni-Shi'i tension and Persia's recognition of Iraq will be discussed. In the following chapter, the negotiations for a settlement of the frontier dispute and the

Treaty of 1937 will be dealt with. The Kurdish problem and the revival of Shiʿi activities following the Iraq Revolution of 1958, will be discussed in subsequent chapters.

A brief account of the Shiʿi community's position in Iraq following the British occupation of the country might be useful for an understanding of the events and tensions that ensued after the establishment of the Arab regime, which adversely affected the relationship between Persia and Iraq. It is difficult to determine accurately what the numerical strength of the Shiʿi community was when the British expeditionary force took control of Iraq. More difficult indeed is to estimate the percentage of Shiʿi followers in the country, as most of the Shiʿa, outside of the major urban centers, were tribesmen, and it would be exceedingly difficult to provide even a rough figure.[6] After the First World War, when Iraq was detached from the Ottoman Empire, the Shiʿi leaders began to claim that they formed the majority of the population. Perhaps such a claim might have been justified were the Kurdish area, whose status had not yet been determined by the peace treaty with Turkey (1924), excluded from the country. Indeed, the Kurds refused to participate in the Arab regime when it was set up in 1921. It is, however, doubtful that the Shiʿa had become the majority in the 1920s, as their numerical strength began to increase in the late 1930s, and they seem to have slightly outnumbered the Sunni community after the Second World War.

It was on the basis of their numerical strength that the Shiʿa began to assert their claim to leadership. But in reality they had never tried to validate that claim by participating constructively in the political processes that had just been made available under the Arab regime. True, they demanded the establishment of an elective national assembly and organized political parties to participate in the political system; but their methods were on the whole uncooperative. Refusing to participate in the general elections for the constituent assembly, the mujtahids issued *fatwas* (legal opinions) to boycott the elections in 1923 and turned down appointments to high political offices on the grounds that the regime was dominated by foreign influence. Thus their denial of the legitimacy of the regime rested on religious grounds, and they resorted to violent acts which aggravated sectarian tensions and led to unrest in the tribal areas of the Middle Euphrates.

The position of the Shiʿi community was compounded by two other factors. First, Shiʿi sanctuaries in the cities of Najaf, Karbala, Samarraʾ; and Kazimayn had become the centers of visitation not only for the Shiʿa of Iraq but also for Shiʿi followers elsewhere, especially in Persia. Moreover, Najaf, where the tomb of ʿAli is to be found, has long been considered a great center of learning, and many mujtahids who distinguished themselves as high spiritual leaders have studied there.

Second, some of the Persian mujtahids who studied in Najaf or other centers of learning, such as Qum and Mashhad, chose to stay in Iraq and retain their Persian nationality, which afforded them greater security under the Ottoman regime. In time their number increased and they intermarried with Shi'i followers of Arab descent. Under the new Iraq regime, the children of Persian nationals born in Iraq could claim both Iraqi citizenship by virtue of their birth in Iraqi territory (*jus soli*) and Persian nationality by descent (*jus sanquine*), although their parents could retain their Persian nationality without adopting Iraqi citizenship. Some of the Persian mujtahids reached the highest rank of spiritual leadership and often issued *fatwas* to Shi'i followers on political questions of the day. For instance, Mirza Taqi al-Shirazi, the mujtahid of Karbala, and Shaykh Abu al-Hasan al-Isfahani, the mujtahid of Najaf, whose *fatwas* proved instrumental in stirring the tribes of the Middle Euphrates during the Iraq revolt of 1920, were Persian nationals. Following their deaths (in 1920), the high religious authority (al-marji'iya) passed to Shaykh Husayn al-Nayini, the mujtahid of Najaf, and Shaykh Mahdi al-Khalisi, the mujtahid of Kazimayn, who were also Persian nationals, although the latter seems to have had a mixed Iraqi (Arab) and Persian descent. Not only did Shaykh Khalisi participate in the Iraq revolt of 1920, he also participated in Shi'i political activities after the establishment of the Iraqi national regime. Other Persian mujtahids, such as Ayat-Allah Kashani, who supported the Mussadiq government (1953), and Ayat-Allah Khumayni, the spiritual leader of the Islamic Revolution (1979), had resided in Najaf before they returned to Tihran and Qum to take an active part in their country's revolution. It was thus that the Shi'i mujtahids, whether in Iraq or Persia, sought to serve the larger interests of the Shi'i community irrespective of geographical segregation or national identity. In their endeavors to serve ultimate Islamic goals, they came into conflict with their compatriots who sought to assert modern concepts such as national (secular) and local interests. Domestic conflicts stemming from sectarian differences strained the relationship between Iraq and Persia, influencing the forces and events that led to the outbreak of war between the two countries.

Following the First World War, when Iraq and Persia had just emerged as nation-states on the new political map of the Middle East, the mujtahids in both countries became very active in their bid for power. In Persia, the mujtahids, through their participation in the nationalist movement, succeeded in reducing British influence and enabled Riza Khan, founder of the Pahlawi (Pahlavi) dynasty, to rise to the throne and replace the Qajar dynasty in 1923. In 1979, the mujtahids led the Islamic Revolution, overthrowing the Pahlawi dynasty. Moreover, the mujtahids in Persia have always tried to lend assistance to their co-religionists in Iraq, and they were often supported by their home government. The mujtahids in Iraq, however, have not always been in a position to support Persian mujtahids, owing to conflicting national interests and Perso-Iraqi relationships were often complicated by sectarian tensions.

King Faysal, known for his caution and moderation, tried to follow a balanced policy after he ascended the throne, and he was not unsympathetic to Shi'i aspirations. Even before Faysal assumed power, Shaykh Mahdi al-Khalisi and a few other Shi'i leaders had sent cables to King Husayn of the Hijaz requesting his approval for the nomination of his son Faysal for the throne of Iraq. Upon his arrival, King Faysal met with Shaykh Khalisi and assured him of his sympathy with Shi'i demands for the country's independence. Further assurances were given to other Shi'i leaders, who seem to have been satisfied that the Arab regime under Faysal would be immune to foreign influences.[7]

Faysal, however, was unable to give satisfaction to all pressure groups. Since he owed his throne to Britain as well as to some of its leading administrators in Iraq, such as High Commissioner Sir Percy Cox and his staff (Kinahan Cornwallis, Gertrude Bell and others), who gave him full support, he was unable to resist foreign pressures when, under instructions from their home government, they advised him that Iraq would have to accept British tutelage until the League of Nations Mandate had been terminated. Moreover, most of the Arab leaders who served under Faysal in Arabia, Syria, or Iraq were Sunnites, especially the established houses that had come into prominence under the Ottoman regime—the Gaylani, Suwaydi, Sa'dun and others. Nor was there an adequate number of educated Shi'i young men to enter government service, as most Shi'i leaders were either spiritual or tribal leaders who had received little or no education. Faysal began to encourage Shi'i followers to enter government (secular) schools and promised them high offices after graduation.

Shi'i leaders on the whole were not unaware of Faysal's difficult position, and some moderate Shi'i leaders were prepared to cooperate with him. But the extremists, including Shaykh Mahdi al-Khalisi, who felt they were betrayed, turned against him. In their attack on British policy, the Shi'i leaders concentrated on specific issues which appeared to limit the country's independence. They also criticized Iraqi (Sunni) leaders who either cooperated with or failed to stand up against the British. They were particularly critical of Sayyid 'Abd al-Rahman al-Gaylani, head of the Sunni community, who became King Faysal's first prime minister. He and the members of his Cabinet came under attack for preparing to conclude a treaty of alliance with Britain which would recognize the British Mandate and compromise the country's independence. King Faysal may not have been very happy with the British proposals for the treaty, but he realized that his country was still in need of British financial assistance and defense against possible foreign attacks.[8]

In April 1922, when the Cabinet was discussing the British proposals for the treaty (to which several ministers were opposed), the southern desert area of the country was invaded by the Ikhwan (Wahhabi Brethren), led by Faysal al-Dawish, a tribal leader from the Saudi desert. The mujtahids, led by Shaykh Mahdi al-Khalisi, seized the occasion and issued an appeal to tribal leaders (both Sunni and Shiʿi) to assemble in Karbala to consider measures for the defense of the country against the Ikhwan. Since the Ikhwan were an extremist Sunni group, the raid aroused Shiʿi feeling and the Shiʿi tribal leaders responded favorably to Shaykh Khalisi's call, but it was disregarded by Sunni tribal leaders, who held that the matter was political and should be resolved by the government. Concerned that the meeting be strictly confined to Shaykh Khalisi's declared object, the government dispatched a force to prevent disturbances that might lead to disorder or to an uprising such as that which had occurred in 1920. Meanwhile, in an exchange of letters between Baghdad and Riyad, the Ikhwan were ordered to withdraw and Ibn Saud, ruler of Saudi Arabia, apologized for the raid. The meeting at Karbala was thus confined to the signing of a petition to the King demanding that measures be taken to defend the country from Ikhawan's raids. The mujtahids immediate object—to display their patriotic concern about British influence—was achieved, as it was rumored that the British had encouraged the Ikhwan's raid in order to put pressure on the Cabinet to accept their proposals for the treaty. But the mujtahid's ultimate objective was to undermine the position of the Cabinet dominated by Sunni leaders and to influence King Faysal to increase the Shiʿi representation in it.[9]

Following the Karbala meeting, the mujtahids, joined by Sunni extremists (for entirely different reasons), continued to agitate against the British mandate and demand that the treaty, still under consideration between Britain and Iraq, should acknowledge Iraq's independence. Rumors were set afoot that, despite opposition by some of the ministers, the Cabinet was under pressure to accept the treaty within the framework of the Mandate. On May 24, 1922, Shaykh Mahdi al-Khalisi circulated a telegram among the tribal Shaykhs of the Middle Euphrates, informing them of British pressure on the Cabinet to accept the treaty, and the telegram was published in one of the leading Baghdad papers.

Meanwhile, the extreme Sunni and Shiʿi leaders demanded permission to organize political parties as a means of mobilizing public opinion against the treaty. Permission to organize parties was granted in August, when the

law of association was enacted in which certain restrictions were provided
to restrain activities that might disturb public order. Three parties applied
for license—Hizb al-Watani (National Party), led by Ja'far Abu al-Timman
(a Shi'i merchant); Hizb al-Nahda (Revival Party), organized by Muhammad
al-Sadr (a mujtahid of Kazimayn); and a moderate party, to counterbalance
the two opposition parties, led by Mahmud al-Gaylani (son of the prime
minister), often referred to as the Government Party. Although stronger in
its number of supporters, it proved ineffective. The other two parties, similar in objectives, merged to form one opposition party.

Even before the opposition party was formed, its members had been agitating in Baghdad and Najaf, and a petition had been addressed to the King
demanding complete independence and rejection of the treaty. It was rumored that the leaders also demanded the resignation of the Cabinet, headed
by Gaylani, and threatened to boycott the elections as a protest against its
policies. Gaylani offered to resign unless King Faysal declared himself in
support of the Cabinet. Not fully satisfied with the terms of the treaty, he
made no explicit declaration in support of the Cabinet and insisted that his
policy had not changed. Thereupon, Gaylani submitted his resignation. The
opposition, welcoming the fall of the Cabinet, hoped that the King would
invite its leader, Muhammad al-Sadr, to form a new Cabinet headed by a
moderate Shi'i leader thus recognizing the strength of the Shi'i position. But
the King was reluctant to take such a step, which would put him on the side
of the opposition against his British supporters and might render his own
position at the mercy of Shi'i leadership.

The crisis was resolved a few days later when the anniversary of the
King's accession to the throne was celebrated on August 24, 1922. Sir Percy
Cox, British High Commissioner, went to the Royal Court to offer his congratulations to the King Faysal. By coincidence, the leaders of the opposition were also there to offer their own congratulations, and they submitted
a petition demanding that a constituent assembly be called to lay down a
constitution; that general elections be held for Parliament; and that a treaty
be signed with Britain which would ensure the country's independence. While
Sir Percy Cox was still at the Court, the King's secretary made a speech in
reply to one that had been made by the opposition. Both speeches were
highly rhetorical, inciting one of the crowd attending the celebration to shout
"Down with the Mandate," to which the others responded with loud ap-
plause. When the tenor of the speeches became known to Sir Percy Cox, he
sent a strong protest to the King on the following day demanding an apol-
ogy, which he received on the following day.

Meanwhile, the King was suffering from a sudden attack of appendicitis,
which his medical officers decided required an immediate operation. For the

next few days, before the King could attend to the business of government, the country was without King and Cabinet, as the Prime Minister had already resigned. Sir Percy Cox seized the opportunity to assume full responsibility. The country may indeed have been in need of tranquillity and security, but Sir Percy Cox took utmost advantage of the situation. He ordered that the principal leaders of the opposition be deported to the Island of Hinjam in the Gulf, suspended operation of opposition parties, and suppressed two of the papers that published news about opposition activities. Moreover, Muhammad al-Sadr and Shaykh Muhammad al-Khalisi, son of Shaykh Mahdi al-Khalisi, were exiled because of their participation in the agitation against the treaty and the British Mandate. Although some disorder occurred in the tribal areas, which was brought under control by the British Air Force, no significant disturbances were reported in urban centers. As a result, the Shi'i leaders lost the battle in their bid for power.

After he resumed authority on September 10, 1922, King Faysal was reported to have been satisfied with the actions of High Commissioner Cox and thanked him for the measures he had taken in a letter addressed to him and made public. Gaylani, who resigned a fortnight earlier, was invited to form a new Cabinet. Only one portfolio—the Ministry of Education—was retained by a Shi'i member, since Ja'far Abu al-Timman, who held the portfolio of commerce, was excluded on the grounds of his activities with opposition leaders. The treaty, which was the subject of controversy, was signed finally on October 10, 1922, after the British Secretary of State had given an assurance to the Iraqi government that the Mandate would be terminated from the moment Iraq was admitted to membership in the League of Nations, and he promised to obtain the admission of Iraq at the earliest possible opportunity.[10]

Before their influence was brought under control, the mujtahids made still another attempt to agitate against the regime. Following the signing of the treaty, the Cabinet resigned on November 16, 1922, and a new Cabinet, headed by 'Abd al-Muhsin al-Sa'dun, was formed with instructions to hold elections for the constituent assembly. In preparation for the elections, the political leaders who had been deported to Hinjam by Sir Percy Cox were released. Moreover, as evidence that Britain was not insisting on a prolonged period of control, the duration of the treaty was cut from twenty years to four. While the King and the government welcomed the action and looked forward to genuine cooperation with Britain, the extremists continued to denounce the British as masters of the country. It was to these extremists and their Shi'i supporters that the mujtahids addressed their appeal to stand against the King and his British supporters.

Early in June, 1923, the mujtahids issued *fatwas* forbidding participation in the elections for the constituent assembly. The *fatwas* were taken as an attack against the government. Since the country was still under rigid control, the freedom necessary for the holding of elections was denied. One incident provided the government a pretext to act. A nephew of Shaykh Mahdi al-Khalisi was arrested on June 21, 1923, while attempting to post a copy of an anti-election *fatwa* on the gate of the Kazimayn Mosque. Sympathizers who were watching the scene tried to secure his release, but a police force was dispatched to take him to the police station. Meanwhile, the son of Shaykh Khalisi also tried to post a copy of the *fatwa;* when the police moved to arrest him a few of his followers rushed to release him, attacking the police. Reinforcements were dispatched to disperse the crowd, and two of Shaykh Khalisi's sons were arrested. The arrests prompted Shaykh Khalisi to call on shopkeepers to close their shops in protest against the government's action, but only a few seem to have closed and no demonstrations took place. While the King was on a visit to the southern provinces, the Cabinet met to discuss Shaykh Khalisi's challenge to its authority. Since the penal code had already been revised earlier in June to give the government the power to deport foreigners for political activities, the Cabinet decided to deport Shaykh Khalisi and his two sons (as well as a nephew) on the grounds of their political activities as foreign (Persian) subjects. They were taken to Basra by a special train and sent by boat to Adan (Aden). From there they went to Makka for a pilgrimage before they made their way back to Persia.

The Persian mujtahids of Najaf protested against the government's action, and the shopkeepers of Najaf closed on June 27, 1923. Under the leadership of two mujtahids—Abu al-Hasan al-Isfahani and Husayn al-Nayini—a large party of mujtahids and their disciples set out for Karbala, hoping to influence others to join in a general exodus of Shi'i mujtahids to Persia. Although a large number was aroused, only fifty, including the two mujtahids, arrived at Karbala on June 29. By order of the Minister of Interior, the governor of Karbala told the mujtahids that the government had no intention of taking action against them if they abstained from interfering in politics. Only a few, hardly more than thirty-five, including students and relatives, were determined to leave. They left the country on July 3, 1923, in a special train to Khanaqin and crossed the frontier to Karmanshah, presumably awaiting the reaction in Iraq. Although the government's action was resented by the Shi'i community, the tribal leaders were not prepared to rise up in arms in their support as they did in 1920. Nor was the Persian government expected to protest Iraq's action. The British Minister in Tihran, to avoid possible reaction in Persia, flew to Baghdad on July 22 to discuss possible solutions. The government, however, was firm against the mujtahid's interference in the elections and assured the British Minister that it would be prepared to allow the mujtahids to return after the constituent assembly had completed its work. Thereupon, the mujtahids proceeded on their way to Qum, the center of religious activities. Two of them decided to return, with the intention of proposing terms for the return of their colleagues; they arrived at Kazimayn on September 12 and returned to their homes in Najaf. In October, Shaykh Mahdi al-Khalisi returned to Persia and went to Qum. After his arrival, a misunderstanding with some who had gone into exile for his sake occurred, and their desire to return home prompted them to contact the Iraqi authorities. Early in 1924, the Shiʿi leaders in Iraq began to intercede on their behalf. The government made it clear that there was no objection to the return of the mujtahids, with the exception of Shykh Mahdi al-Khalisi, provided that they refrain from interfering in politics. They were allowed to return in April of 1924. Shaykh Mahdi al-Khalisi, who went to reside in Mashhad, continued to denounce both the Iraqi and the British governments.[11]

After the elections for the constituent assembly were over and the work of the assembly—the ratification of the Anglo-Iraq Treaty, promulgation of the Constitution, and enactment of an electoral law—was completed, King Faysal began to take a conciliatory approach to the Shiʿi community. Even before the assembly was dissolved (1924), Prime Minister Saʿdun resigned in November 1923, partly because of Shiʿi resentment toward his Cabinet, whose differences with Shaykh Mahdi al-Khalisi were irreconcilable, and partly because of differences between the King and Premier on administrative matters. A new Cabinet, headed by Jaʿfar al-ʿAskari, was formed, in which two portfolios—Finance and Education—were given to Shiʿi leaders, denoting a favorable change of attitude toward the Shiʿi community. As a consequence, a deputation of Shiʿi leaders presented a letter to the King in which they expressed loyalty to the Crown and cooperation with the government. As a demonstration of their loyalty, the Shiʿi leaders expressed their support of the transfer of the Caliphate from the Ottoman House to the Royal House of Husayn, father of Faysal, when the Turkish government abolished the Caliphate in 1924.[12] Meanwhile, tension was diffused when the King paid an official visit to Karbala on the occasion of the opening of a railway extension, and then proceeded to Najaf on pilgrimage. In both cities he was warmly received.

While relations between the Shiʿi community in Iraq and the government began to improve, relations between Persia and Iraq remained cool. Persia refused to recognize the Iraq government, and official diplomatic relations were conducted through the British diplomatic mission in Tihran, which passed its messages to Iraq through the British High Commissioner in Baghdad, although Persia enjoyed consular privileges in Iraq. Since this one-sided relationship between two neighbors was unusual, with privileges accorded by one side to the other, it was bound to be brought to the attention of the League of Nations. In its *Report on the Administration of Iraq* for the year 1926, the British Government stated:

> It is noteworthy that Persia is the only member of the League of Nations having close relations with Iraq and maintaining consular officers in the country, which has so far failed to accord formal recognition . . . the Persian Government appears to consider that their refusal to recognize Iraq prohibits them from co-operating in certain measures which would be beneficial to both countries, such as the establishment of direct telegraphic communication and of a combined customs port at the frontier, on the main trade route between the two countries. The absence of these two facilities contribute, to no small extent, to the hampering of commercial relations.[13]

In its examination of this report first by the Permanent Mandates Commission of the League of Nations and then by one of the committees of the League's Assembly, which discuss matters relating to countries under the Mandate, Persia's refusal to recognize the Iraqi government was noted. In the Assembly's Committee, Muhammad ʿAli Khan Furughi (Foroughi), the Persian delegate, stated in his comment on Persia's relations with Iraq that the Iraq authorities had mistreated Persian nationals living in that country.[14] At a meeting of the Permanent Mandates Commission, M. Rappard, a member of the committee, said that he was surprised that Persia, refusing to recognize Iraq, had consulates in Iraq. In reply to Rappard's comment, Furuguhi said that "Persia does not keep any consuls officially appointed as such in Iraq, but only agents in charge of consulates."[15] With regard to Persian nationals resident in Iraq, he claimed that they were entitled to special judicial privileges granted to certain foreigners in accordance with the Anglo-Iraq Treaty of 1922, because they were badly treated in the Iraqi courts.

B. H. Bourdillon, British accredited representative, in reply to both the Persian delegate to the League and to Rappard admitted that certain "difficulties" had existed between Iraq and Persia, but he hoped that a solution could "be found by friendly discussion." With regard to Rappard's question about the Persian consuls, Bourdillon said that a letter dated November 7, 1927, from the Persian Foreign Office to the British minister in Tihran might illustrate the manner in which the appointment was made:

> The Persian Ministry for Foreign Affairs begs to bring to the notice of the British Legation the fact that Mirza Javad Khan Vahid, ex-Persian Consul at Erivan, has been appointed Persian cousul at Mosul and has left for his destination.[16]

Bourdillon went on to explain that the procedure followed by the British minister in Tihran was to inform the British high commissioner in Baghdad of the appointment. The high commissioner in turn informed the Iraq Government and requested that obstacles not be put in the way of the fulfillment of such an appointment, presumably on the grounds that the post was not a new creation, as it had existed in Ottoman times. In pursuing this formality, Persia continued to keep its consulates in Iraq without extending recognition to it. The Iraq government, however, allowed Persian consuls to use their full titles and to carry out all the normal functions of consular officers.

Nonetheless, Iraq continued to press for recognition, as Persia's demand for equality of treatment with foreigners under the Anglo-Iraq treaty was not as simple as Persia claimed. Not all Persian nationals resident in Iraq were foreigners. There were many Perisans who had been born in Iraq—some even had parents who had been born in Iraq—and who could claim Persian nationality on the grounds that the Persian law of nationality recognized persons born of Persian parents as Persian nationals, irrespective of the territory in which they were born. The Iraqi nationality law, on the other hand, based partly on the Treaty of Lausanne (1924) and partly on general principles recognized by other nations, states that a person born in Iraq is an Iraqi national if his father was also born in Iraq. As a result, a large number of Persians possessed dual nationality. In general, the rule is that the nationality of the country of residence prevails. Thus a Persian national born in Iraq would be treated as a Iraqi and in Persia as a Persian. However, in neither country could he claim the consular protection of the other country. For these reasons, Persia was not accorded the same privileges for its nationals as other countries, in accordance with the Anglo-Iraq judicial agreement, yet Persia continued to claim the right to protect persons who were, under the Iraqi law, Iraqi nationals and who had not taken steps to renounce their Persian nationality.

The Persian claim, based on the principle of equal treatment, lost much of its force when the British government "informed the Iraq government that in its opinion a British national who is also an Iraqi national is not entitled to special privileges."[17] So long as the Anglo-Iraq judicial agreement was in force, however, Persia insisted on its rights to equality and refused to recognize Iraq.

But this was not all. Persia continued to complain through the British high commissioner in Baghdad about the mistreatment of its nationals, whether in Iraqi courts or in other social and business relationships.[18] The Persian delegate to the League of Nations echoed these complaints. He went so far as to assert that the Iraqi judicial system was inadequate to ensure justice and that Persian nationals were discriminated on confessional (Shiʿi) grounds, notwithstanding that often Shiʿi judges were on the bench in connection with suits relating to Shi'i followers. Moreover, the accredited British representative stated before the Permanent Mandates Commission that the newly established system of justice in Iraq was far better than that which had existed under the Ottoman regime.[19]

Since Persia insisted on its right to equal treatment, despite offers to negotiate the differences, its refusal to recognize Iraq continued until 1929 when Britain and Iraq finally agreed to terminate the judicial agreement and abolish the special privileges granted to foreigners. In March 1929, the British delegate informed the Council of the League of Nations that the "speical privileges" would be abolished, to be replaced by a uniform system of justice in Iraq. The proposal was adopted, provided the countries enjoying the special privileges were willing to renounce them, and Britain proceeded to carry out the proposal. A new judicial agreement was concluded with Iraq, which provided for the retention of a British legal judge in the Iraqi judiciary within the uniform system. Meanwhile, negotiations with countries enjoying special privileges, including the United States, proceeded, with mutual satisfaction, to end the legacy of the Ottoman system of capitulations.[20]

No sooner had the League of Nations approved the British proposal to abolish the "special privileges" than Persia and Iraq moved quickly to initiate direct negotiations and establish normal diplomatic relations in 1929. The Shah of Persia and the King of Iraq exchanged telegrams (April 3, 1929) in which the Shah congratulated the King for the League's approval of the abolition of the judicial agreement and removal of "the obstacles that lay between them." The King of Iraq, in his reply, thanked him for his good wished and added that he looked forward to "the restoration of the means of stable and friendly relations between two neighboring nations which are bound by strong and old established ties of fraternity." The King sent his private secretary—Rustum Haydar, a Shi'i follower himself—to Tihran to convey his personal appreciation of the Shah's telegram on April 20, and it was agreed to exchange notes of recognition in August 1929. These exchanges prompted Persia to appoint Mirza Taqi Khan Nabawi as its first Minister to Iraq in February 1930, and Iraq subsequently appointed Tawfiq al-Suwaydi, a former premier, as its first Minister to Persia in March 1931. From that time on, Persia and Iraq began not only normal diplomatic relations, but also negotiations to resolve other longstanding issues.[21]

# IV

# The Iraq-Iran Conflict during
# the Inter-War Years:
# The Diplomatic and Legal
# Aspects

From the moment Iraq was formally recognized by Persia in 1929, relations between the two countries were conducted directly by their governments. But before Iraq became independent, and thus no longer under the British Mandate, in 1932, Britain continued to provide guidance for the conduct of Iraq's foreign relations and was consulted on all major questions between the two neighbors. Iraq had to deal with several pending issues in its relationship with Persia, such as dual nationality, demarcation of frontiers, and the migration of tribes across borders. Britain often rendered invaluable assistance by throwing light on issues that it had already discussed with Persia on behalf of Iraq.

The status of persons possessing dual nationality was one of the perennial problems that had faced the authorities in both countries ever since the Iraqi nationality law was enacted in 1924, owing to the existence of a large number of Persian nationals resident in Iraq. It was estimated that over 250,000 people possessed dual nationality; they were Iraqis under the Iraqi nationality law and Persians under the Persian nationality law. According to the Iraqi nationality law, those who chose to preserve the nationality of their ancestors were permitted to do so within a year from the promulgation of the law.[1] Problems arising from the absence of an official census and from the delay in registration because of people's ignorance of their status led to complaints by Persia that its nationals had been mistreated. The problem was compounded when some of the Persian nationals were not of Persian descent. A case in point is the Muhaysin tribesmen in the Basra Province, who were Arabs living on both sides of the Shatt al-Arab. The Iraqi govern-

ment maintained that the tribesmen living on the right bank, in Iraqi territory, where they had been in residence for many generations, were Iraqi
subjects. Because of fear of the possibility of conscription, some seem to
have held Persian nationality and did not present themselves before the census officials for registration. Thus several persons had been prosecuted for
one irregularity or another and were either fined or imprisoned. This and
other, similar situations confronted by town-dwellers and tribesmen living
on both sides of the River Euphrates, especially in such cities and towns as
Basra and Muhammara (later Khurumshahr), were the cause of almost continual friction and misunderstanding between the Persian and Iraqi authorities.[2] Although these problems were indirectly dealt with through the British
High Commissioner in Baghdad, settlement of most of them by peaceful
methods was possible even before the Iraqi government was recognized by
Persia in 1929. After the establishment of formal diplomatic relations, the
major issues centered on the crossing of frontiers by tribes and rebels seeking asylum—incidents that both sides tried to resolve by negotiations. Recurrence of these incidents have not been completely stopped, partly because
of family or tribal relationships of people living on both sides of the frontier,
and partly because of the incomplete boundary demarcations on land since
Ottoman times, which led to disputes that both Iraq and Persia have yet to
resolve.

After 1932, when the British Mandate came to an end and Iraq was recognized as an independent state, the conduct of foreign relations became
Iraq's responsibility. Iraq was no longer under obligation to consult Britain,
save on matters of foreign policy and the defence of the country in accordance with the Anglo-Iraq Treaty of 1930. In domestic affairs, the only Iraqi
administrative unit in which Britain retained an interest was the Basra Port,
which had its own budget because it owed the British government a debt to
be paid annually. As long as Britain was still involved in the Basra Port,
Persia seems to have exercised self-restraint, although its claim to participate
in the navigation in the Shatt al-Arab was asserted. When British influence
in Iraq began to recede after independence, the questions of navigation in
the Shatt al-Arab, in which the Basra Port was involved, and of the whole
frontier line between Basra and the Gulf were raised by Persia for reconsi-
deration. But before we address ourselves to the larger frontier dispute, perhaps a few words about the Basra Port would be useful at this stage.

The Basra Port, constructed by Britain during the First World War, was
administered as a separate unit long before it passed under Iraqi jurisdiction.
Before the First World War, there were no port facilities in Basra, and
foreign cargo vessels could not be loaded or unloaded except by lighter

vessels. The occupation of Basra by British forces in 1914 and the deployment of troops required the introduction of modern equipment and port services. In 1919, the military functions of the Basra Port were no longer needed, and the port began to be administered by the British authorities on a commercial basis.

On the establishment of the Iraqi national regime in 1921, the Basra Port was administered as a civil organization by a port director and an advisory committee, composed of British and Iraqi personnel. Moreover, the Port Trust was set up in accordance with the Financial Agreement (subsidiary to the Anglo-Iraq Treaty of 1922), to deal with the budget as well as with the port as a debt to be paid to Britain, which had undertaken the initial construction of the port. Although the Port Trust was not established to deal with the question of valuation, the budget was excluded from the general account of the Iraqi government, and the terms of payment of the debt were negotiated by a separate agreement. This working arrangement lasted until British tutelage was superseded by Iraq's rise to independence in accordance with the Treaty of 1930, which provided for the transfer of responsibility for the port to Iraq when the debt was finally paid.[3]

During the entire period under British tutelage, the services rendered by the Basra Port to all foreign vessels seem to have been satisfactory to all foreign countries, including Persia, whose vessels carryied cargo to Persian ports on the Shatt al-Arab. The Anglo-Persian Oil Company, which operated from Abadan in shipping oil through the Shatt al-Arab to other countries, cooperated with the Basra Port authorities in providing loans for dredging and maintaining a deep-water channel. No serious problem seems to have arisen before Iraq assumed full responsibility for the Port, but the situation began gradually to change when Persia put forth claims to the right of navigation and other matters in the early 1930s.

Following the First World War, both Persia and Iraq emerged as new nation-states. Both began to conduct their foreign policies in accordance with the principles and practices of modern diplomacy and International Law. Neither country entertained the idea of resorting to war or using force to settle their differences between them, and they agreed to abide by the League of Nations Covenant and the Treaty of Paris (1928), which renounced war as an instrument of national policy, save in self-defence. These commitments were taken very seriously when both Iraq and Persia agreed to conduct themselves as modern nation-states. Thus when Iraq was admitted into the League of Nations (in 1932), both countries sought to cultivate friendly relations and cooperate through the new world organization to which they belonged. King Faysal, aware of his country's constraints and limited resources, had no ambition to embark on foreign ventures. Above all, by his good-neighbor policy, the King sought to win the friendship of Persia and resolve longstanding issues by peaceful methods.

Persia, however, looked at its relationship with Iraq from a different perspective. True, the Shah reciprocated the friendship and cooperation offered by King Faysal, but he and his ministers sought by personal contacts and formal relationships to gain certain national advantages in their negotiations with the Iraqi regime, especially in their claim to the Shatt al-Arab and other frontier issues. From the time that Iraq was recognized as an independent state, the Persian and Iraqi governments became unable to resolve the conflicting claims over the boundary line of Shatt al-Arab by direct negotiations. King Faysal and his ministers were made aware of this situation even before Iraq achieved independence. The boundary question was made clear to them when King Faysal paid a state visit to Tihran in 1931, a year before Iraq gained its independence. During that visit, accompanied by Prime Minister Nuri al-Saʿid, King Faysal was cordially welcomed by the Shah, who placed the Gulistan Palace at his disposal, and both sides displayed the highest respect for each other. But when both the King and his first minister had occasion to discuss frontier problems, the Shah and his ministers told their Iraqi guests that they expected them to concede certain national demands as a gesture of friendship. Even before King Faysal visited Iran, the British High Commissioner in Baghdad warned him about the Persian demands, since he had visited Tihran in early April and discussed informally, among other things, the frontier problem with the Persian authorities. When the question was raised directly with King Faysal, he asked Prime Minister Nuri to discuss the matter after his departure. The Prime Minister, unable to come to an understanding with the Persian ministers, confined his negotiations to less complex questions. Negotiations concerning frontiers and navigation in the Shatt al-Arab were postponed first for a month and then indefinitely for several years.[4]

On the occasion of the admission of Iraq to membership in the League of Nations in 1932, the Persian delegate, while joining other League members in welcoming Iraq's admission, remarked that there were still a number of threaties to be negotiated and other border issues to be resolved. The formal relations between the two countries remained cordial, but friction and increasing frontier violations prompted the foreign offices of the two countries to exchange notes and protests that aroused the concern of higher authorities on both sides. It is perhaps unneccessary to provide a chronology of frontier incidents and violations, but the frequency with which they occurred convinced both sides that unless the larger question of the frontier was resolved, the frontier incidents would continue unabated and if violence were resorted to, it would be exceedingly difficult to bring under control.[5]

Direct negotiations led nowhere because Persia contested Iraq's claim to sovereignty over the whole of Shatt al-Arab and demanded the thalweg, the navigable line of the river (ordinarily applicable to international rivers), to be the frontier line. It was decided to bring the dispute to the attention of the Council of the League of Nations, as both countries were members of that organization. Iraq, which took the initiative, submitted the dispute in a letter dated November 29, 1934, and Persia gave its consent in a note dated December 23, 1934.[6]

Since Iraq's case rested on several treaties that Persia had entered into with the Ottoman Empire, perhaps a brief review of the treaties relevant to the frontier problem would be illuminating. We have already seen how the Ottoman Empire and Persia, after a long period of rivalry and warfare, finally agreed that the maintenance of peace was in their mutual interest, despite the confessional (sectarian) differences that existed between them. The peaceful coexistence of two neighbors, however, necessarily raised the question of the need for frontiers to separate the territorial jurisdiction of one country from the other. More than a dozen treaties that would establish permanent frontiers between them and put an end to differences were concluded. These treaties may be divided into two categories.

The first category, consisting of over half a dozen instruments, may be described as armistice agreements that stopped fighting between the Ottoman and Persian forces, but could not establish peace or lay down permanent frontiers. True, under the Treaty of Zuhab (1639), the first important attempt to make peace, the two countries sought to identify the provinces that belonged to each side, but no clear boundary line was drawn to separate one territory from the other, as such frontier lines were considered to separate the territory of Islam (dar al-Islam) from the territory of unbelievers, called the territory of war (dar al-harb) according to the Islamic law of nations.[7] For this reason, Persia and the Ottoman Empire called the Treaty of Zuhab a sulh (truce) and not silm (peace), since the latter meant permanent conditions of peace among believers, while the former was a short span of peace with the unbelievers—a truce that should not exceed ten years.[8] Indeed, neither the Ottoman Sultan nor the Persian Shah intended that the treaty concluded between them last too long, since both were anxious to resume fighting at the earliest possible moment in order to reestablish the unity of the house of Islam under one supreme authority. Only under the rule of a Caliph or an Imam (depending on who would be enthroned to rule) did believers expect peace to endure. But when they finally realized that peace and unity could not be achieved by war, the Persian and Ottoman authorities decided to enter into agreements to establish peace without an attempt to impose one of the two creeds—Sunnism or Shi'ism—on each other. The initiative to conclude a treaty embodying the principles of unity and peace was first undertaken by Nadir Shah (d. 1747), who proposed to reduce Shi'ism into a fifth school of law, called the Ja'fari school, provided that Sunni scholars would recognize Shi'ism within the Islamic superstructure. Before he submitted to the Ottoman Sultan the draft treaty calling for reconciliation, Nadir Shah held a conference at Najaf in 1743, to which Shi'i and Sunni scholars were invited, at which his proposal for a Sunni–Shi'i compromise was approved.[9] The treaty was signed in 1746 at Kurdan (a town near Tihran) by both Nadir Shah and the Ottoman Sultan's accredited representative. Before the Sultan would ratify the treaty, as we noted before, he sounded the opinion of Sunni scholars, who counseled against the recognition of Shi'ism as a fifth school of law. For this reason, the Sultan, while accepting the principles of peace and territorial segregation, rejected the principle of reconciliation, just as European princes, rejecting the unity of the Christian creed had agreed to establish peace under the Treaty of Westphalia (1648), under which religious differences were relegated to the domestic plane.

The second category of treaties dealt more specifically with frontier questions and reaffirmed the principles of peace, territorial sovereignty, and the relegation of confessional differences to the domestic plane, which had tacitly been accepted under earlier treaties. This category of treaties, consisting of some half a dozen instruments—treaties, protocols, and *procès-verbaux*—dealt with almost all frontier problems. The most important instruments were two treaties: the so called first and second treaties of Arzurum (Ard al-Rum). The first was concluded on July 28, 1823, the second, on May 31, 1847.[10]

The first Treaty of Arzurum may be regarded as a step to the second, since it reaffirmed the areas belonging to the Ottoman Empire and Persia as laid down in the Zuhab (1639) and Kurdan (1746) treaties, defining frontier zones rather than frontier lines separating the territories of the two countries. Although the Kurdan Treaty had not been ratified by the Sultan, it is significant that reference to it was made, including the principles of Sunni–Shi'i reconcilation and the normalization of Perso-Ottoman relationships on the basis of peace and exchange of permanent diplomatic missions. "The Two High Powers," the treaty added, "do not admit each other's interference in the internal affairs of their respective states" (Article 1). It also states that "a Minister shall be sent every three years to reside for that period at the respective Courts" (Article 7). Persian pilgrims were allowed to travel without obstructions whether they visited the holy places in Makka and Madina or in Najaf and Karbala, and Persian merchants had to pay the duties which all subjects of the sultan had to pay. The terms of the treaty were not respected; the Ottoman forces attacked the Persian town of Muhammara on Shatt al-Arab in 1837, and the Persian forces entered Sulaymaniya (a principal center in Kurdistan) in 1840, each claiming the occupied locality to be under its jurisdiction. Thus the peace treaty of 1823 was breached, and the mutual respect for a clear frontier line was still needed. This need was felt not only by the two sides, but also by Britain and Russia; these two great powers had become deeply involved in Ottoman and Persian affairs, and the maintenance of peace in the region became necessary if the great powers were to avoid wars arising from regional conflicts. Good offices were offered and accepted, and the representatives of the four states met at Arzurum to conclude a second treaty bearing the same name in 1847.

The second Treaty of Arzurum addressed itself specifically to frontier problems. Its provisions may be summarized as follows:

1. Persia ceded the "lowlands" situated to the west of the province of Zuhab, and the Ottoman Empire ceded the "mountainous" eastern part of that province (including the Kirind Valley).
2. Persia abandoned its claim to the city and province of Sulaymaniya and promised not to interfere in the Kurdish affair of that province.
3. The Ottoman Empire recognized Persian sovereignty over the city and port of Mahammara, the island of Khizr, and the anchorage and land on the eastern bank of Shatt al-Arab, which were in the possession of tribes considered to be Persian subjects.
4. Persian vessels had the right to navigate freely on the Shatt al-Arab from the head of the Gulf to the point of contact of the frontier of the two countries.
5. Persia and the Ottoman government agreed to appoint commissioners and engineers as representatives for the purpose of demarcation of the frontier agreed on under the treaty. Moreover, special commissioners were appointed to decide on all cases of damage resulting from frontier settlement submitted by either party.
6. Persian pilgrims were granted the right to visit the holy places in the Ottoman dominions, as before.
7. The two parties agreed to suppress brigandage and other acts of aggression committed by tribesmen settled on both sides of the frontier.
8. Consuls were to be stationed in centers deemed necessary to render services to nationals of one party in the territory of the other on the basis of reciprocity.

The two Arzurum treaties, taken together, may be said to have provided the framework for a peace settlement perhaps more meaningful than any other that the Ottoman Empire and Persia had entered into before. The three basic principles of peace, territorial sovereignty, and noninterference in domestic affairs had finally been agreed on for the normalization of relations between the two Muslim states.

As evidence of good faith, the two Muslim states agreed to establish permanent frontier landmarks, exchange diplomatic missions, and grant their nationals privileges to carry out business and visit holy places and other activities in their respective territories without obstruction. Mixed commissions, including British and Russian representatives, were appointed for the demarcation of frontiers. Differences arising from the exercise of reciprocal rights or the carrying out of any of the functions specified under the treaties would be resolved by negotiations or any other peaceful method, in accordance with the terms of the two treaties.

Nevertheless, the implementation of the treaties gave rise to almost innumerable incidents, and differences over the interpretation of the articles relating to frontiers delayed the demarcation for over half a century. Indeed, by the time the ottoman empire became involved in the First World War and was subsequently dismembered, the demarcation commission had not finished its task and a relatively small portion in the southern area had yet to be completed. When Iraq came into existence as a successor state, it inherited not only the problems arising from the unfulfilled task of the demarcation commission, but also the legacy of the historic differences between Persia and the Ottoman Empire. Some of the problems, especially those relating to frontiers, arose from differences in the interpretation of several clauses of the second Treaty of Arzurum. Others resulted from the new conditions created after the First World War, when both Iraq and Persia began to reconsider their relationship as new nation-states in accordance with standards entirely different from those followed in the Perso-Ottoman era.

The differences that arose as a result of the conflicting interpretations of the treaty may be summarized as follows: (1) some of the clauses that appeared either too vague or too general gave an opportunity to one side to provide a different interpretation from the other; (2) since the demarcation commission had not been granted the power to decide on matters arising from delimitation, differences were bound to be referred to higher authorities, which slowed down the process of work on the spot considerably. Despite agreement to resolve issues in two protocols, one signed in Tihran in 1911 and the other in Constantinople in 1913—the first provided for the referral of differences to an arbitration tribunal at The Hague, and the other resolved issues supplementary to the Treaty of Arzurum—the work of the demarcation commission was not completed when the First World War broke out in 1914.[11]

Following the war, when negotiations were resumed, Persia sought to review the whole frontier issue anew, not only the frontier incidents and violations, on the grounds that the validity of the Arzurum Treaty and the protocols was in question. It was Persia's refusal to settle the incidents and violations within the framework of the Arzurum Treaty and the protocols that prompted Iraq to refer the matter to the League of Nations in 1934.

Iraq's complaints against Persia rested essentially on Persia's refusal to recognize frontiers based on agreements that it had already accepted in its negotiations with the Ottoman Empire. But there were other specific matters, such as Persia's interference with navigation in the Shatt al-Arab, which was under the control of the Basra Port directorate; its establishment of police posts on Iraqi territory; its claim of ownership over Sarakushk, a strip of territory in the central boundary; and its construction of dams on the Gunjam River that diverted water essential for irrigation in Iraq. These and other matters were submitted for discussion at the forthcoming meeting of the Council of the League of Nations.[12]

In stating Iraq's case before the League Council, General Nuri al-Saʿid, foreign minister of Iraq, asserted his country's rights on the grounds of equity no less than on treaty arrangements. He said,

On the general question of equity, the Iraqi Government feels that it is Iraq and not Persia that has grounds for complaints. Persia had a coast-line of almost two thousands kilometers, with many ports and anchorages. In the Khor Musa, only fifty kilometers away to the east of the Shatt al-Arab, Persia possesses a deep-water harbor penetrating far into Persian territory, where she has already constructed the terminus of the Trans-Persian Railway. Iraq is essentially the land of the two rivers, Euphrates and Tigris. The Shatt al-Arab, formed by their junction, constitutes Iraq's only access to the sea; it required constant attention if it is to be kept fit for navigation by modern shipping, and Basra, 100 kilometers from the mouth, is Iraq's only port. It is highly undesirable, from Iraq's point of view, that another Power should command this channel from one bank. Iraq is not asking that the frontier be altered, but I make these remarks to show that this is not because the existing line is unduly to its disadvantage.[13]

General Nuri was on more firm grounds when he argued that Iraq's claim to sovereignty over the whole Shatt al-Arab was based on the Second Treaty of Arzurum and on the protocols of 1911 and 1913. The latter agreements provided for the formation of a four-power commission to draw up the frontier in detail, and the commission had proceeded to the region to delimit and mark the boundary on the spot. General Nuri went on to explain: "The

Commission completed the whole of its work, except for one small sector north of Mount Dalampar and therefore outside the area with which we are concerned. . . . It is this clear and well-considered boundary which my Government wishes to see respected as the frontier between the two countries.''[14]

As to Persia's claim that the thalweg should be the only rule to be applied to international rivers, General Nuri rightly stated that it is by no means a universal rule. He might have added that the Shatt al-Arab, as it then existed, was not strictly speaking an international river. ''The boundary can be,'' General Nuri added, ''and sometimes is, fixed at the bank by agreement, and when this is the case, the agreement is incontestably valid.''[15]

On January 15, 1935, Baqir Khan Kazimi, Persian Foreign Minister, in stating his country's case, rejected the claim of Iraq to sovereignty over the Shatt al-Arab on the grounds that the Treaty of Arzurum made no reference to sovereignty over the Shatt al-Arab; that the Constantinople Protocol, although not the Tihran Protocol, had not been ratified; and that the Persian representative who had appeared at Constantinople had exceeded his powers in accepting the Constantinople Protocol when it was signed by the Ottoman authorities. Kazimi had already submitted a memorandum, on January 8, 1935, in which he set out the views of his country in detail. He summarized the reasons for rejecting the protocols of 1913 and 1914, as follows:

> They must be rejected, (1) because they take as a starting point a treaty which was itself non-existent at the time when the Tihran Agreement of 1911 referred to it; (2) because, in concluding the 1913 Protocol which already gravely departs from the provisions and stipulations of the Tihran Agreement of 1911 providing for arbitration in case of disagreement . . . , (3) because, on the pretext of a treaty between Persia and the Ottoman Empire, accompanied on the Shatt al-Arab by a direct agreement concluded in London between Great Britain and the Sublime Porte and by the improper conclusion of a bilateral understanding in the British capital in the middle of negotiations which were to take place at Constantinople between all parties; (4) because, lastly, one of the Parties, the Ottoman Empire, immediately failed to carry out, in a great many respects, the Act of 1913 fixing the frontier—a failure which, even if partial, involved the total lapse of the Act owing to its indivisible character.[16]

It is clear from the Persian and Iraqi arguments that the position of the former rested essentially on procedural and not always on substantial evidence. For instance, Persia rejected the Constantinople Protocol of 1911 and the record of the commission of 1914, but accept the validity of the Tihran Protocol of 1913, even though the other documents were negotiated and signed according to the same procedure followed at Tihran. Moreover, Persia raised the constitutional question of ratification to declare that its treaties with the Ottoman Empire were not binding after they had been signed. Under International Law, once a treaty is signed by competent authorities, it becomes binding, as constitutional methods vary from country to country and heads of state are expected to be aware of their constitutional limitations when they sign or authorize others to sign treaties. Nor is the assertion that international rivers must conform to the thalweg rule unequivocally acceptable to most jurists. But Persia seems to have overstressed this procedural agreement to declare its treaty obligations not binding long after they had been signed and resorted to bringing pressure to bear on Iraq to accept its demands.

The Council of the League of Nations, realizing that these divergent legal agruments could not be easily reconciled, suggested further direct negotiations and appointed a rapporteur, Baron Aloisi, the Italian delegate, to offer his good offices when the negotiations were resumed. At first, Aloisi suggested that the internationalization of the Shatt al-Arab might be the answer to the question. But his suggestion was acceptable neither to Iraq nor to Iran (renamed in 1935). Iraq proposed to refer the dispute to the Court of International Justice for an advisory opinion, but Iran rejected the offer. The negotiation, however, was removed from the agenda of the League early in 1936 by agreement of the two parties. The scene shifted from Geneva to Tihran, and General Nuri continued direct negotiations with the Iranian higher authorities. When the negotiations reached a point of almost breaking down, the Shah himself gave further encouragement when, in a private audience early in 1937, he hinted to General Nuri that his country wanted nothing more from Iraq than the thalweg of the Shatt al-Arab in front of Abadan. This meant that Iraq would retain its sovereignty over the whole Shatt al-Arab except for a few kilometers of the Abadan area, and Nuri accepted this compromise as a basis for further negotiations.

Complications in the international situation, prompted by Mussolini's ambition to extend his control to the eastern Mediterranean and his occupation of Ethiopia in 1935, alerted not only Iraq and Iran, but also Turkey and Afganistan to the common threat to the region beyond the eastern Mediterranean; they began at once to contemplate the formation of a Middle East security pact. Turkey was the first to feel the danger of Mussolini's threat and therefore suggested that Iran and Iraq speedily settle their differences in order to discuss the larger problem of regional security. The president of Turkey cabled a friendly personal message to the rulers of Iraq and Iran, expressing the hope of a satisfactory agreement. In these circumstances, Iran and Iraq were ready to be more flexible and accepted the compromise that led to the signing, on July 8, 1937, of the Sa'dabad Pact, a regional-security agreement of four Middle Eastern countries—Turkey, Iran, Iraq, and Afganistan. This pact was the impetus that inspired Iran and Iraq to negotiate a new treaty governing navigation in the Shatt al-Arab.[17]

While the delegates of the four Middle Eastern powers were meeting to sign the Saʿdabad Pact in Tihran, the two principal delegates of Iran and Iraq met separately to sign the boundary treaty (July 4, 1937), which was expected to settle the longstanding dispute between the two countries. The treaty confirmed the validity of the Constantinople Protocol of 1913 and the *procès-verbal* of the delimitation commission of 1914 as bases for the revised frontier between Iraq and Iran (Article 1). It also provided that the frontier between the two countries would run along the Shatt al-Arab on its left bank, except for the section of eight kilometers in front of Abadan, where the frontier would be the thalweg of the Shatt al-Arab (Article 2). The two countries agreed to set up a commission to erect "frontier marks," determined by the commission, as evidence of the frontier line between the two countries (Article 3). It was also agreed that the two countries would conclude a convention to deal with all questions concerning navigation in the Shatt al-Arab, such as dredging, pilotage, collection of dues and, other matters relating to the improvement of the navigable channel (Article 5).

In a protocol attached to the treaty, Iraq and Iran agreed that the authorization given by one party to

a warship or other public vessel used for non-commercial purposes belonging to a third state to enter ports belonging to said . . . Party and situated on the Shatt al-Arab shall be regarded as having been given by the other . . . Party in order that such vessel may make use of its waters when passing through the Shatt al-Arab. . . . [And the] Party who has given such an authorization must inform the other Party thereof immediately.

The protocol also provided that "nothing in this treaty prejudices the rights of Iraq and its obligations undertaken towards the British Government regarding the Shatt al-Arab in accordance with Article 4 of the treaty dated June 30, 1930." [18]

The 1937 treaty, which was expected to resolve longstanding issues and remove obstacles that hindered the two countries from cooperating for their common interests, suffered from certain unclarified matters that rendered implementation exceedingly difficult. Moreover, the circumstances, both domestic and foreign, in which the treaty had been born began to change and affected adversely the attitude of the two countries toward each other. Even before the treaty was signed, the government of Iraq, whose foreign minister, General Nuri, had been instrumental in the negotiation of the treaty, was overthrown by a military coup d'état. Frequent changes in Iraqi regimes gave Iran the opportunity to put forth new demands when means of implementation were discussed. Above all, the new conditions created by the Second World War and the postwar years proved unfavorable for implementation, and frequent violations of the treaty by one side or the other brought the two countries almost to the brink of war.

# V

# The Revolutionary Movement in Iraq and Iran's Reaction to It

We have already seen how Iraq and Iran sought under their old regimes to resolve by peaceful means the long standing issues they had inherited from the past and to develop fairly good neighborly relations. True, the treaty of 1937, although settling basic frontier issues, left much to be desired in its specification of details, especially concerning the supervision of navigation in the Shatt al-Arab waters. But the tide of confessional tension seemed to have subsided; Shi'i followers in Iraq, including those of Persian descent, had become reconciled to the Iraqi regime, and their leaders were able to make their way up into high political offices. As the inter-war years drew to a close, Iraq and Iran were cooperating in promoting their common interests and advocating peace and security through the Sa'dabad Pact and bilateral arrangements. It seemed that the two countries were heading for a long era of amicable relation.

But the prospects of peace in the world at large were not very promising. On September 1, 1939, hardly two years after the two neighbors had ironed out their differences, the Second World War broke out. The Gulf region, including Iraq and Iran, were drawn into the war. Both countries endured military occupations that lasted from 1941 to 1946, when the last foreign forces finally withdrew. In the aftermath of the war, conditions in both countries began to change considerably, and each began to advocate a particular line of social and political development that affected adversely the foreign policy they had pursued before the war.

In Iran, the founder of the Pahlavi dynasty, Shah Riza Khan, abdicated in 1941 in favor of his son, Shah Muhammad Riza. For almost a decade, the son sought to consolidate his regime and pursue the policy of modernization that his father had laid down. Lacking the charisma and experience of his father, he might have been able to achieve modernization by democratic processes because the Majlis (Parliament) had just begun to exercise

42

its powers after his father's abdication and would have provided validation and perhaps public support for his policies. But his ambition to govern the country in accordance with his father's high-handed rule defeated his endeavors to achieve quick and meaningful reforms. More specifically, in foreign policy, his father had been able to play off the two historic rival powers—the Soviet Union and Britain—against each other and maintain the country's independence. But the American support that Muhammad Riza solicited for his imperial design resulted in the gradual alienation of the people and the weakening of his regime. The purpose of American policy makers was to encourage the Shah to pursue a constructive domestic policy and not to replace British with American influence, but the Shah used the American support to enhance his own power and authoritarian rule, which was resented by the people. Nor did his policy, as subsequent events demonstrated, serve the best interests of the West. From a regional perspective, the Shah's imperial policy was viewed with suspicion and disfavor in Arab lands, and his support for the Kurdish War in Iraq and the expulsion of Shiʿi mujtahids from his country aroused the concern of both Iraq and other Arab Gulf countries.

In Iraq, the domestic and foreign policies under the old regime, not unlike Iran's policies under the old Shah, consisted in the main of promoting internal reform and development and asserting the country's independence and good-neighborly relationships. These policies, reputed to have been shaped by King Faysal I (d. 1933) and pursued by General Nuri al-Saʿid after him, proved quite adequate for Iraq's requirements. After the Second World War, however, conditions began to change. The income from oil, made available for development, appeared to serve vested interests since it was earmarked for largely long-term projects, and short-term projects that might have improved general conditions were sadly neglected. Above all, General Nuri's foreign policy, faulted because he had committed the country to a Western alliance, was opposed by the people who demanded neutrality in the conflict between East and West. Nor were the new generation and its supporters satisfied with Nuri's domestic policy, as they were denied the right to participate in public affairs. As in Iran, an uneasy alliance eventually developed among dissatisfied groups, irrespective of their differences, and it swept away the old regime. The Iraqi army (especially the young officers), unlike the Iranian army (which by tradition was loyal to the monarchy and sought to protect the old regime, even against popular protests, until it was overthrown by the Islamic Revolution in 1979), shared the grievances of the new generation and came into conflict with the monarchy. After several attempts, the army officers, calling themselves free officers—free from loyalty to the old regime—seized the power and overthrew the regime in July 1958.

The Shah's attitude toward the July Revolution was, not unnaturally, averse, and he was at first unwilling even to recognize the new Iraqi regime, hoping that it might be overthrown by foreign intervention. When, however, the revolutionary regime, presided over by Brigadier 'Abd al-Karim Qasim, was recognized by several countries, including the Western powers, the Shah reluctantly followed suit and recognized it two weeks later, since it pledged to respect international agreements and cooperate not only with Arab neighbors, but also with all Muslim states.

The Shah, nevertheless, continued to be suspicious of Iraq's military leaders, and the tension between the two countries remained high. He and his royal entourage, it was reported, were horrified by the manner in which the members of the royal house of Iraq were executed and by the brutality of killing by an angry mob spurred by mass hysteria. Perhaps even more important was the overthrow of the civilian regime by a military revolution. It alarmed the Shah as well as other civilian rulers in the region, since the wave of military coups that had begun in Egypt in 1952 was showing signs of spreading into other Arab countries. It set an example for revolutionary leaders, military or otherwise, in other countries.

Nor was the ideological tide in the Arab world less alarming to the Shah. Indeed, Iran had already experienced the danger of ideological propaganda that invited Soviet intervention, and not without difficulty did the central authority escape being overthrown by the Iranian Communist party. True, the new regime in Baghdad was not in the hands of Communists, but Qasim had allowed the Communist party to infiltrate into several government departments and influence his policy. Moreover, he entered into several economic and military agreements with the Soviet Union that encouraged Communists to spread propaganda and incite uprisings that might have invited Soviet intervention. The Shah had already been alarmed by the increasing Soviet influence in Syria and Egypt under Nasir. So he had reason to be apprehensive of the trends of events in Iraq and the possibility of joining other Arab lands in the anti-Western wave that appeared to encircle and isolate his country from the West. Small wonder that the Shah was exceedingly reluctant to deal with Qasim and come to an understanding with his regime.

Even before the July Revolution erupted in Baghdad, the Shah had keenly felt that only the United States could challenge the Soviet ambition to penetrate into the gulf region. But the United States, although responding favorably to the Shah's overtures of cooperation, moved too slowly to commit itself to play the role of an ally. Only after Qasim had concluded an agreement with the Soviet Union in 1959 and received economic and military assistance did the United States agree to enter into slightly more comprehensive agreements and promise to assist Iran were it to become the subject of a foreign attack.[1] Thus the United States stopped short of acting as an ally, as the Shah had hoped, perhaps mainly because it did not want to go beyond the position taken by the soviet Union in its bilateral agreements with Iraq and provoke Qasim to enter into a formal alliance with the Soviet Union.[2] But although the Shah's expectations were not completely fulfilled, he was prepared to proceed slowly in engaging the United States to enhance his position against increasing Soviet influence in Arab lands. Despite American assurance of support, the Shah considered the military regime established in Baghdad by revolution unworthy of his trust, and he sought to undermine Qasim's regime and perhaps to replace it with a government more agreeable to him. To achieve this goal, the Shah began to revive grievances and issues that had existed in the past between Iran and Iraq.

First, there still existed the historic boundary conflict over the waters of Shatt al-Arab. Under the Treaty of 1937, the boundary between the two countries, which had been the eastern bank of the river, was slightly modified by granting Iran an area (roughly about eight kilometers) in front of ʿAbadan, defined by the thalweg line. It was also agreed that a joint commission should be established to deal with navigation problems in the river on the basis of a convention to be concluded between Iraq and Iran. Because of disagreement about the nature and functions of the joint commission, Iraq was reluctant to conclude such a convention and undertook to supervise navigation without regard to Iranian complaints. Under the old regime, several attempts were made in vain to resolve the issue, but no agreement was reached.

At the outset, the Shah tried to learn Qasim's position on the question of navigation. In the fall of 1958, shortly after the Qasim regime was recognized by Iran, the Iranian delegation to the United Nations discussed the subject with the Iraqi delegation, but no significant progress was reached on the subject. Indeed, the situation became even worse when Iraq decided to withdraw from the Baghdad Pact in 1959, and Iranian shipping in the Shatt al-Arab was confronted with further obstructions. Sharp notes were exchanged on the subject, and the Shah, in one of his press conferences (November 28, 1959), criticized Iraq for refusing to settle the navigation question by peaceful methods. Qasim, angered by the Shah's criticism, denounced the Shah's claims and reasserted Iraq's sovereignty over the entire river without regard to the provision of the 1937 treaty concerning the eight-kilometer thalweg line in front of ʿAbadan. Despite the exchange of sharp statements, neither country undertook hostile acts against the other, although the tension adversely affected trade relations between them and visitations by Iranians to Shiʿi holy places in Iraq were interrupted.

Second, in 1961 Qasim became involved in another problem that he suspected to have incited by the Shah—the Kurdish War. The Kurds in Iraq, although fewer in number than the Kurds in Iran, have always been more vocal in expressing their national aspirations and had several times in the past revolted against the central authority. The Kurdish problem became even more difficult to deal with after the Second World War, partly because of Soviet instigation but mainly because Kurdish nationalism came into sharp conflict with the rising tide of Pan-Arabism (a problem about which more will be said later). Under the Qasim regime, the Kurds formally put forth the demand for autonomy. Qasim, although at first making many vague promises, reneged and failed to understand the depth of Kurdish nationalism.

The Kurds, divided among four neighbors—Turkey, Iran, Iraq, and Syria—did not really want independence; they aspired to have a form of autonomy within Iraq that would allow them to manage their own local affairs. Mulla Mustafa of Barzan, one of the vigilant tribal leaders, championed the cause of autonomy and emerged as the most formidable leader in the postwar years. He left Iraq to participate in the establishment of the Kurdish Republic at Mahabad in 1946, and he was given the military rank of general. After the collapse of the Mahabad Republic, he went with a few followers to the Soviet Union, in 1947, where he seems to have been given some military training, since he was unable to return to Iraq.

After the fall of the old regime in 1958, Mulla Mustafa, whose national reputation had been enhanced, returned to Baghdad. He paid several visits to Qasim, and an understanding on cooperation between Arabs and Kurds seems to have been reached. Qasim sought Kurdish support for his regime, but he did not fulfill his promises. The young Kurdish generation expected greater freedom under the new regime. When Qasim showed no interest in their aspirations, they were bound to follow Mulla Mustafa's leadership, even though he clung to traditional tribal leadership. Qasim, becoming aware of Mulla Mustafa's ambition, began to restrict his activities when he began to tour the Kurdish areas early in 1959 and to rally support for his movement. The rift between Mulla Mustafa and Qasim gradually deepened and in 1961 led to war, which lasted for more than two years.

The Shah must have had mixed feelings about the Kurdish War in Iraq. For even if the Shah were not prepared to lend the Kurds his support, the war would undermine Qasim's position and might lead to the downfall of his regime. But Mulla Mustafa's self-exile in the Soviet Union and the support he had received from liberal and Communist groups after his return to Baghdad gave the impression that the Kurdish movement in Iraq was the ally of Communism. Moreover, Mulla Mustafa's visit to Moscow in 1960 at the invitation of the Soviet Union, for the anniversary of the October Revolution (he returned to Iraq in March 1961), seems to have aroused the Shah's suspicion and gave him the impression that the Kurdish War had been instigated by the Soviet Union. In these circumstances, he was not prepared to provide assistance to Mulla Mustafa, although Iranian Kurds, who sympathized with their Iraqi brethren, gave indirect support. The failure of Mulla Mustafa to receive direct support from the Shah prompted him to come to an understanding with ʿAbd al-Salam ʿArif, who succeeded Qasim in 1963. He recognized the futility of the prolongation of a war without substantial foreign support. When Mulla Mustafa resumed the war to achieve his cherished ambition of Kurdish autonomy under the Baʿth regime, the Shah lost no time in offering his support. We shall return to this subject later.[3]

Third, still another source of trouble for Qasim's regime was the religious leadership. Although the Iraqi clerics were supported by Iranian sympathizers, the Shah was not at heart in sympathy with them, save their opposition to Qasim. These elements, both Sunnis and Shiʿis, were formally organized when Qasim allowed political parties to be licensed shortly after his regime was established. Like the Kurds, the Shah was reluctant to deal with groups to which he was opposed in his own country, as they were considered a threat to his regime. But when Qasim expelled Shiʿi followers of Persian descent in retaliation against Iran's claims to the thalweg as the border line in Shatt al-Arab, the Iranian government could not be indifferent to the Iraqi action. Beyond condemnation by the Iranian Majlis and exchange of rhetorical denunciations by the press in both countries, no serious step was undertaken against Iraq by the Iranian government.

But the mujtahids were not idle. They called on Shiʿi followers in Iraq to criticize the Qasim regime and organized demonstrations to protest its actions. Muhsin al-Hakim, the leading Shiʿi religious leader in Najaf, became very active in Iraqi politics. When two religious groups submitted applications to the authorities to allow them to operate as licensed parties, al-Hakim's name was put up as a sponsor of one of them. This group was obviously an anti-Communist party, since it stated in in its program that it aimed at combating atheism and secularism. The Minister of the Interior, doubtful of the fidelity of its leaders, denied them permission to organize. On appeal, the court decided in favor of the founders, and the Islamic party began to operate in 1960.

Hakim and his followers did not make a frontal attack against the regime, and their criticism was rather mild at first. Later, when their criticism of the Communist party, became more outspoken, presuming that Qasim was sympathetic to Communists, it was taken to be against the regime. On Febru-

ary 12, 1960, Hakim issued a *fatwa* (legal opinion) in reply to a question put
to him about whether it was permissible under Islamic law to join the Communist party, whose teaching stressed "disbelief and atheism." The text of
the *fatwa*, published in the press, was welcomed in many circles, Sunni and
Shiʿi, and considered implicitly hostile to the regime. Later, Hakim spoke
his mind more openly when the Islamic party issued a proclamation, on July
5, 1960, in which the people were warned against Communist malicious
propaganda and activities. Three months later, on October 15, 1960, the
Islamic party submitted a petition to Qasim in which the government was
criticized in very strong terms for neglecting religious instruction and espousing Communist and atheist teachings. More specifically, the petition
demanded that Communism be abolished, that all publications and papers
preaching the Communist creed be suppressed, and that religious leaders
who had been arrested or thrown into prison during the past two years be
released. As a result of the increasing criticism, the license of the Islamic
party was withdrawn in 1961, and its leaders were arrested on the grounds
of stirring unrest and hostile propaganda against the regime. Religious opposition to Communist activities contributed in no small measure to undermine the Qasim regime, to the great satisfaction of the Shah; but the Shah's
support to Hakim was indirect because Shiʿi religious leaders warned Hakim
that the Shah was opposed to the mujtahids in Iran. A scion of Hakim,
Muhammad Baqir al-Hakim, became a supporter of Ayat-Allah Khumayni,
the future spiritual leader of the Iranian Revolution, when Khumayni arrived
at Najaf in 1965. We shall return to the younger Hakim's activities.

From the fall of Qasim, in 1963, to the coming of the Baʿth party to
power, in 1968, the strained relations between Iran and Iraq were considerably reduced, mainly because the regimes that followed that of Qasim—first
under ʿAbd al-Salam ʿArif and then under his brother ʿAbd al-Rahman ʿArif—
appeared anti-Communist to the Shah. No major issue seems to have been
raised by the ʿArif brothers, and the two problems arising from the construction of an oil pipeline by Iraq and the question of sovereignty in the off-
shore waters of the Gulf were dealt with by negotiations.

In April 1963, the Iraqi government decided to construct an oil pipeline
(twelve-inch capacity) from Khanaqin (near the Iraq-Iran border) to Baghdad. The oil reserves of both Khanaqin, on the Iraqi side, and the Nafti-
Shah and Khana, on the Iranian side, seem to have the same subterranean
source. The Iranian government argued that the Iraqi action would increase
the exploitation of the common oil reserve at the expense of Iran. It was the
first time since the fall of the monarchy in Iraq that the two countries agreed
to resolve pending issues by peaceful means. In July 1963, ʿAbd al-ʿAziz
al-Wattari, the Iraqi Minister of Oil, visited Tihran, and an agreement was

reached on the annual volume of oil that each country could produce and the method of supervision (each nation was empowered to inspect the operation of the other).

With regard to the Gulf off-shore waters, in which Iraq claimed that Iran had infringed its territorial rights, it was agreed to resolve the issue on a "joint exploitation" basis, but how to do so was left for discussion in a subsequent meeting. When, however, the Iraqi Minister of Foreign Affairs, Subhi ʿAbd al-Hamid, visited Tihran in February 1964, no significant progress was achieved, and the subject was left for further negotiation. Both sides seem to have avoided carrying on negotiations that might have disturbed the tranquillity between them. Revealing anti-Communist propensities and a friendly attitude toward the West, Iraq under the two ʿArif brothers appeared no longer an enemy to the Shah. He seems to have temporarily put aside the differences that he had raised with his next-door neighbor on minor issues. But this situation did not last very long.

The rise to power of the Baʿth party, which advocated a blend of Pan-Arab ideas and socialism, was an unwelcome event to the Shah. The Baʿth party seized power by a military coup, and it advocated a nonaligned foreign policy whose beneficiary, in the Shah's eyes, was none other than the Soviet Union. The new regime in Baghdad was thus considered by the Shah to be opposed not only to Iran, but also to the West as a whole.

At the outset, the Shah hoped that the Baʿth regime, not unlike that of Qasim, would be short-lived and might soon be replaced by another, more moderate and accommodating to him. During the first two years of Baʿth rule, he noted that the Baʿth leaders were so preoccupied with the struggle for power with their opponents that the regime seemed unstable and might be replaced by a more moderate one at any moment. He even went so far as to support two retired army officers, Major General ʿAbd al-Ghani al-Rawi and Colonel Salih Mahdi al-Samarraʾi—the first a former supporter of the ʿArif brothers' regimes, and the other a protégé of the old regime—who made a vain attempt to overthrow the regime by a military coup d'état on January 20, 1970. Their secret contact with Iran, discovered by the Iraqi police, led to the arrest of the conspirators red-handed, and they were brought to trial. The principal leaders, Samarraʾi and others, were put to death, but General Rawi, who managed to stay out of the country, was sentenced *in absentia*.

Suspecting Iran's complicity in the abortive coup, the Baʿth government on January 22, 1970, ordered the Iranian ambassador, ʿIzzat-Allah ʿAmili, and four members of his staff to leave the country within twenty-four hours.[4] The staff of the Iranian consulates in Baghdad, Basra, and Karbala were expelled at the same time. Iran retaliated within a few hours by ordering the expulsion of the Iraqi ambassador in Tihran, the military attaché, and four members of his staff. Tension between the two countries rose so high that war was expected to break out at any moment, as Iran moved its forces to the Iraqi border. This action prompted Iraq to appeal to the United Nations, on February 2, to take measures to prevent the situation from degenerating into actual fighting. Since Iraq had no intention of going to war with Iran, the Iraqi Minister of the Interior, Salih Mahdi ʿAmmash, visited Ankara on February 3, requesting Turkish good offices to prevent the crisis from developing into an armed conflict. Informed of the talk between the Iraqi minister and the Turkish government, Iran replied that if Iraq were to attack, Iranian troops would fight to defend their country, and it proposed that both sides withdraw their troops from the frontier. War was averted, but tension and suspicion between the two regimes continued.

Even before he tried to replace the Baʿth regime by means of a counter-coup, the Shah had raised the question of navigation in the Shatt al-Arab, shortly after the Baʿth party seized power in 1968. Navigation in the river was a longstanding issue that the Shah had raised on more than one occasion, especially during his confrontation with the Qasim regime, but the issue remained unresolved. Because the Shah was especially opposed to the Baʿth Pan-Arab policy, which asserted Arab unity and regarded the Shatt al-Arab as the eastern boundary of the Arab homeland, the dispute over navigation was revived on both legal and political grounds. The Baʿth regime, however, was determined to protect Iraq's territorial integrity on the strength of the 1937 treaty and was not prepared to capitulate to Iranian demands. Since the Shah had failed to pressure Iraq by political machinations, he sought by legal and diplomatic means to achieve his objective.

International rivers, it is true, have always raised technical and administrative problems, such as changes in the river course and variations in the deep channel, for riparian states. But the Shatt al-Arab is only in part shared with Iran and may not be considered, strictly speaking, an international river.[5] Navigation in the Shatt al-Arab had become a bone of contention mainly because Iran claimed that the 1937 treaty left the matter to be dealt with by both countries in a subsequent convention that would establish a joint commission whose functions and terms of reference were not defined, although that treaty made no mention of a joint commission to supervise navigation (Articles 5). After the Second World War and Britain's withdrawal of its forces from Iraq, the Shah began to make demands that exceeded the provisions of the 1937 treaty. We have already discussed the Shah's maneuvers in raising the issue on legal and technical grounds, demanding that the frontier question be reconsidered as a whole before he denounced the treaty in 1969. Rejecting Iran's demand, the Baʿth government insisted that any dispute arising from the 1937 treaty be settled in accordance with the methods provided under the accord attached to it, which stipulates that all boundary differences between the two countries be settled by either arbitration or judicial procedure.[6] But the Shah seems to have abandoned the legal method, presumably because recourse to judicial settlement would have been unfavorable to Iran.[7]

On April 19, 1969, Iran unilaterally declared that the 1937 treaty was null and void. Since negotiations, even under the benign ʿArif regimes, had not led to agreement, no settlement was expected to materialize under the more assertive Baʿth regime. Before Iran took the drastic step of denunciation, however, an Iranian delegation proceeded to Baghdad in March 1969, and at a meeting with a Iraqi representatives at the Foreign Office, it presented a new draft treaty to replace the 1937 treaty and a protocol entrusting navigation in the Shatt al-Arab to a joint commission. Taken by surprise, the Iraqi delegation pointed out that its functions were confined to technical matters, and it had neither instructions nor power to negotiate on the basis of a new draft treaty.[8] The Iranians, unable to persuade the Iraqis to discuss the new draft, returned to Tirhan empty handed. Meanwhile, Iran continued to assert its control over Iranian shipping by an escort of military vessels, which prompted Iraq to complain that this action was a violation of the 1937 treaty. On April 15, 1969, the Iranian ambassador in Baghdad was told the tenor of the Iraqi complaints in strong terms, which was taken by Iran as a threat and prompted its deputy foreign minister to make a statement in the Majlis in which he declared that the treaty of 1937 was unilaterally denounced, and a note containing the denunciation of the treaty was communicated to Iraq.[9] The Iranian action was followed by accusations and counteraccusations echoed in the press and media of Tihran and Baghdad that aggravated the tension between the two countries.

In these circumstances, Iraq instructed its permanent representative to the United Nations to bring to the attention of the Security Council Iran's action and to submit all the relevant documents concerning the dispute between the two countries. Likewise, the Iranian representative submitted the relevant documents concerning the termination of the 1937 treaty. He reiterated Iran's earlier objection to the treaty—that it was based on Ottoman treaties that Iran had never recognized as valid—and he went on to argue that its provisions had been violated by Iraq for many years. More specifically, he complained about the obstructions imposed on Iranian shipping in the Shatt al-Arab by the Iraqi authorities and Iraq's refusal to cooperate with the Iranian authorities to supervise and improve navigation, although supervision and improvement of the river navigation was Iraq's responsibility in accordance with Clause 2 of the protocol to the 1937 treaty. Moreover, he maintained that the boundary, in accordance with the treaty of 1937, was a line that varied with the rise and fall of the tide, whereas the actual frontier should be a thalweg, as in all other international rivers and in accordance with the general rule of International Law. For these and other reasons (on which the Iranian documents dwelt), he invoked the principle *rebus sic stantibus,* denoting that the provisions of the 1937 treaty were no longer relevant to the new conditions created by the postwar years.

Iraq, considering the treaty of 1937 still binding, declared that Iran's unilateral abrogation of it was contrary to International Law. Indeed, one may argue that there are cogent reasons in favor of Iraq's rejection of Iran's action. Under International Law, a treaty cannot be terminated by one party without the consent of the other, unless the treaty itself specifies such a right as well as the mode of termination. Moreover, Iraq's alleged violations of the treaty provision concerning navigation were due in part to the failure of both parties to agreement to conclude the convention necessary for the establishment of a commission for navigation.[10] Nor does the *clausula rebus sic stantibus* justify Iran's action of denunciation without recourse to judicial procedure, including the consent of Iraq or the opinion (not to speak of the decision) of the International Court of Justice.[11] Of all treaties, boundary treaties are intended to be more enduring than others, unless radical territorial alterations occur that call for revision or change; but such changes, as Brierly rightly pointed out, should not be an excuse for avoiding treaty obligations that a state finds inconvenient to fulfill.[12] For this reason, the 1937 treaty could not possibly be so lightly disregarded, and Iraq demanded that the matter be referred to the International Court of Justice for an advisory opinion. But Iran would have nothing to do with judicial procedure, as the law was on Iraq's side. Failing to achieve his aim by legal means, the Shah reverted to pressure and political machination—support for the Iraqi Kurds in their drive to achieve local autonomy.

It was not the first time that the Kurds had been at war with the central government, and the Kurdish question had become a pawn in the diplomatic game between Iraq and Iran. Since the Kurdish leadership had allowed itself to be exploited by foreign powers and resorted to a war whose beneficiary was none other than Iran, perhaps a brief summary of the background and ramifications of the question might throw light on the conflict between the two neighbors.

Even before it came into power, the Ba'th party had made known its proposals for a settlement of the Kurdish question and its vision of how Arabs and Kurds could live together in peace under a Pan-Arab union in which the Kurds would enjoy an autonomous status. After protracted negotiations between Ba'thist and Kurdish leaders, a memorandum of understanding called "Manifesto on the Peaceful Settlement of the Kurdish Issue," dated March 11, 1970, was proclaimed. It promised that the Kurds would be granted self-rule, to be exercised by a local administrative council and an elected legislative assembly. Adequate guarantees were provided to recognize the Kurdish language as officially coequal with Arabic in the Kurdish district and to promote Kurdish culture and traditions.[13] This plan came nearest to the formula advocated by Kurdish intellectuals, reputed to have been attributed to Jalal al-Tilbani (Talabani), who said that the Kurds would demand semisovereign statehood within an Arab federal union were Iraq to join an Arab union, but would accept autonomy and Iraqi identity if the country remained independent.[14] Since the Ba'th was committed to Pan-Arab unity, the March manifesto provided a happy compromise between Arab and Kurdish national aspirations. But the Ba'th compromise was not destined to materialize because of substantial no less than personal disagreement among Kurdish leaders. Above all, suspicion and distrust that had long been in the making since the creation of the Iraq state frustrated both Arabs and Kurds.

After the First World War, Kurdistan was divided among four states— Turkey, Persia, Iraq, and Syria. The Kurds in Iraq contended that in a country under British influence, they might enjoy greater freedom and their national life might develop into maturity, since the arrangement under the Treaty of Sèvres to establish a larger Kurdish state in eastern Turkey came to naught.[15] Ever since a large portion of Kurdistan has been incorporated into Iraq, the Kurds in Iraq were recognized as a people having their own cultural identity and granted full status as Iraqi citizens. It was hoped that in time, Kurds and Arabs—indeed, all ethnocultural groups—would be integrated to form the Iraqi nation.

After independence, however, neither the Kurds nor the Arabs were prepared to give up their national identify, and the occasional upsurge of Pan-Arabism by the new generation aroused Kurdish suspicion that their dependence on the Iraqi identity might be merely a step toward ultimate assimilation by the Arabs of Iraq. The Kurdish national identity, which had been expected to be superseded by the new Iraqi identity, began to grow and was given an impetus by the rising tide of Pan-Arabism after the Second World War, without a serious attempt to discourage either trend or to impress on both Kurds and Arabs the need for stressing the supremacy of the Iraqi national identity. Nor did the Iraqi Revolution of 1958, which promised to reconcile Kurdish and Arab national aspirations, succeed in allaying Kurdish suspicion. Thus the Kurds went to war with the Qasim regime to achieve their newly proclaimed demand for autonomy. After coming to power in 1968, the Ba'th party made another attempt to resolve the Kurdish question on the basis of proposals noted earlier in the March Manifesto. Although the Kurdish leaders were divided on the matter, an increasing number of younger men were prepared to compromise with the Ba'th regime and accepted self-rule as embodied in the March Manifesto. Mulla Mustafa, however, rejected the Ba'th offer and insisted on autonomy as he understood it.

What prompted Mulla Mustafa to reject the March Manifesto and consequently go to war with the Ba'th government? From his past experience with the Qasim regime in 1961 to 1963—not to mention his earlier experiences— Mulla Mustafa realized that war with the central government, regardless of how long it might last, could not be decisive in obliging Iraq to accept his formula of autonomy without foreign assistance. Thus he had to make up his mind whether he would accept the Ba'th compromise or to seek foreign help.

There were four potential sources of support for the Kurds—the Soviet Union, Iran, the United States, and Israel. In the early postwar years, Mulla Mustafa was convinced that the Soviet Union was perhaps the only great power willing to support his people in their struggle to achieve their national aspirations, since the United States had already been committed to support both Iran and Iraq against Communist penetration into their lands. When the Mahabad Republic was established in 1946, he joined the Kurdish leaders who journeyed to the new center of Kurdish nationalist activities and hoped to stir from there Kurdish national aspirations in Iraqi Kurdistan. After the fall of Mahabad, he went to the Soviet Union, where he remained as an exiled guest for the next twelve years. On his return to Baghdad in 1958, he hoped that the newly established revolutionary regime under Qasim might grant the Kurds national freedom. Very soon, however, Mulla Mustafa became disappointed with Qasim and came into conflict with him. He returned to the Soviet Union in 1961, presumably to attend the celebration of the October Revolution, but in reality he sought Soviet support in his conflict with Qasim. He returned empty-handed, since Qasim had already entered into an agreement with the Soviet Union, and Mulla Mustafa was advised to patch up his differences with Qasim. Nevertheless, he went to war without regard to the Soviet offer of good offices, since he had already come to the conclusion that the Soviet Unon was not interested in Kurdish national aspirations. His disappointment with Soviet reluctance to assist him was perhaps the most important reason for his coming to terms with the 'Arif regime, following Qasim's fall in 1963, and he accepted the proposals for a settlement offered by that regime in 1966. He never again turned to the Soviet Union for counsel or assistance.

The second possible source of assistance was the Shah of Iran. Although Mulla Mustafa received support from compatriots in Iran, he knew that the

Shah had no great love for the Kurds in Iraq, since he had fought the Kurds in Iran and had destroyed the Mahabad Republic; he was prepared to support them only as rebels against the Baʿth regime, not as patriots who aspired to establish a Kurdish national home in Iraq. Moreover, not all Kurdish leaders, especially the younger leaders, were prepared to assist Mulla Mustafa against the Iraqi government, which had promised to improve social and economic conditions in the Kurdish area. Indeed, some Kurdish leaders had begun to cooperate with the Baʿth government before the war broke out in 1974. Mulla Mustafa's own son, ʿUbayd-Allah, deserted him and sided with the government soon after the March Manifesto was issued. ʿAziz ʿAkrawi, one of his close supporters, declared that the faction of the Kurdish Democratic party under his leadership would remain loyal to the central authority. ʿAkrawi was supported by a number of other Kurdish leaders, including Hashim Hasan, then chief of the Executive Council of the Kurdish Province, who held that Kurdish interests would be best served by peaceful rather than by violent methods.[16]

Finally, Israel offered assistance to the Kurds long before the Shah offered his support. Israel had always been ready to deal with Mulla Mustafa and continued to offer its support after he went to war with the central government. Its purpose, shorn of ideological motivation, seems to have been merely to undermine the Baʿth regime because of its outspoken views about the Arab-Israeli conflict. Moreover, since the Baʿth regime refused to recognize Israel and put an end to the state of war between the two countries, Israel sought to cultivate Kurdish friendship as a way to oppose the Iraqi government. Although Mulla Mustafa seems to have accepted Israeli assistance on more than one occasion, he did not go to war with the central authorities when the Iraqi forces were dispatched to participate in the Arab-Israeli war of 1973.[17] It was thus unlikely that Mulla Mustafa would easily fall into the Shah's trap or accept Israeli assistance without first obtaining American assurances that the Shah's promises would be honored.

The Shah, aware that an alliance with Mulla Mustafa could not be easily forged, appealed to the United States for good offices. Since the United States had already been committed to the Shah's policy of playing the role of policeman in the Gulf, American policy makers responded favorably to his appeal and Secretary of State Henry Kissinger assured Mulla Mustafa that the Shah would honor his promises to him in his drive to achieve autonomy for the Kurds.[18]

But how could this uneasy alliance be carried out? It was contemplated that it could be achieved by two methods. First, by military operations in the Kurdish area that would force the Baʿth government to concede Kurdish demands for autonomy as Mulla Mustafa understood it. Second, since the

Shah did not intend to enter the war openly, he resorted to political machination by encouraging elements opposed to the Baʿth to cooperate with Mulla Mustafa and form an Arab-Kurdish front that would achieve power and settle pending issues to the satisfaction of all parties concerned.

Neither method was destined to work. The Baʿth government, which had its own plan for settlement of the Kurdish problem, proceeded to carry it out. On March 11, 1974, the fourth anniversary of the proclamation of the March Manifesto, the government announced that it would enforce the provisions concerning Kurdish self-rule without further delay. Mulla Mustafa rejected it and went to war with the regime. Both sides were counting on a quick victory. During the spring and early summer of 1974, the Iraqi army concentrated on relieving besieged garrisons, opening roads, and moving slowly into the Kurdish area. In July and August, the Pesh Merga, Mulla Mustafa's military forces, was pushed into the mountains along the Turkish and Iranian borders. From this time, the Pesh Merga had to rely on Iranian reinforcements, without which it could not resist the heavy Iraqi offensive, including the bombing of Kurdish towns and villages. For the next six months, the advance of the Iraqi forces into the mountain area slowed down, especially during the winter of 1974. In 1975, the offensive was resumed, but the prospect of a quick victory by either side became uncertain.

Mulla Mustafa's Pesh Merga was confined to the mountains, and an Iraqi offensive in the approaching warm weather would be impossible to stop without substantial Iranian reinforcements. The Shah reluctantly began to have second thoughts about the prospect of a military victory. Nor did the political maneuvering to unseat the Baʿth leadership prove to have any prospect of success. Likewise, a quick military victory by the Iraqis was not imminent. Since the Soviet Union, then the only source of weaponry for Iraq, had stopped supplying needed weapons—not even spare parts—because it had declared its neutrality in the Kurdish War, Iraq's supplies seem to have dwindled considerably; an offensive in the spring of 1975 could not possibly be decisive. Whether the Shah was fully aware that Iraq might not be able to continue the war is difficult to determine, but he seems to have realized that the war was stalemated and a change of regime in Baghdad was not expected.[19] In these circumstances, both the Shah and the Baʿth government must have become mentally prepared to accept the offer of a mediator and come to the negotiating table. Thus the stage was set for a peaceful settlement.

# VI

# A Short-Lived Compromise:
# The Treaty of 1975

Even before Iraq and Iran were ready to meet at the negotiating table, attempts at coming to an understanding by direct negotiations between the two countries had been made, but no meeting of the minds seems to have been reached.[1] In the latter part of 1974, Turkey and Jordan offered their good offices, which may have indirectly paved the way for bringing the two sides together. It was, however, shortly before the closing summit meeting of the Organization of Petroleum Exporting countries (OPEC) in Algiers (March 1975) that Saddam Husayn, Vice President of the Revolutionary Command Council of Iraq, agreed to meet with the Shah of Iran at the invitation of President Hawari Bumidian (Boumedienne) of Algeria, to iron out the differences between the two countries.

Before he left for Algiers, the Shah made no public announcement about his departure, nor did he inform his ministers. They were informed from the plane only after he had left the country. He knew exactly what he wanted in his forthcoming talk with Saddam Husayn, and he was determined that, if an agreement were ever to be reached, his goal must be achieved. Saddam Husayn, on the contrary, had been instructed by his government to insist on the minimum Iraqi security requirements, but on the question of the river frontier, he seems to have been given freedom to compromise with Iran's demands (that is, the thalweg as the border). The outcome of the negotiations thus was in large measure dependent on the judgment of the two leaders, who, after a careful review of the matter, decided to come to a final agreement and to open a new chapter of good-neighborly relations between their countries.

Two meetings, attended by only the Shah, Saddam Husayn, and Bumidian, were held on June 5 and 6, 1975. Bumidian acted as a translator.[2] No agreement was reached on the first day: the Shah insisted on the thalweg as the boundary line in the Shatt al-Arab, which he considered essential for

Iranian security as well as for navigation in the waters of the Shatt al-Arab, and offered to stop his assistance to the Kurds. When the leaders met again on the second day, however, further talks seem to have resulted in an agreement on essential matters of common concern. The agreement centered on the following matters: first, the thalweg would be the boundary line in the Shatt al-Arab; second, the Shah agreed to stop his assistance to the Kurds, which virtually meant that Iraq would deal with the Kurds in accordance with the autonomy proposals offered to them earlier; third, Iraq and Iran agreed to cooperate on the maintenance of peace and security by putting an end to infiltration and subversive activities on both sides of the frontier.

"We agreed," said the Shah, "to bury our differences and succeeded in ending the misunderstandings which the colonialist had maintained between us."[3] The agreement seemed to be a happy compromise that would end all differences, at least as long as the Shah was in power. "The happiness of Iraq," the Shah told Saddam Husayn, "was important to the security of Iran."[4] From the Shah's perspective, the agreement was certainly gratifying, as it achieved his long-cherished demand that the thalweg be the boundary line in the Shatt al-Arab, whereby Iraq would surrender to Iran its sovereignty over half the river. Peace and security, however, proved a mirage, as subsequent events have demonstrated.

Following the agreement between the two leaders, the foreign ministers of the two countries met in separate conferences, which ᶜAbd al-ᶜAziz Butafliqa (Boutaflika), Foreign Minister of Algeria, attended. All matters of mutual concern between the two countries were thoroughly reviewed and embodied in an agreement, which came to be known as the Algiers Agreement, signed on March 6, 1975. The agreement, based on the principles of territorial integrity, the inviolability of borders, and noninterference in internal affairs, provided that the two countries would undertake

1. To make a definitive demarcation of their land frontier in accordance with the Constantinople Protocol of 1913 and the minutes of the then Delimitation of Frontiers Commission of 1914.
2. To define their maritime frontier in accordance with the thalweg.
3. To restore security and mutual trust along their common frontier and to establish strict and effective control that would put an end to all acts of infiltration of a subversive character from either side.
4. To consider the arrangements referred to above as integral elements of the comprehensive settlement. Any violation of the component parts would be contrary to the spirit of the agreement.
5. To reestablish traditional ties of good-neighborly relations and friend-

ship, and to continue to exchange of views on all questions of mutual interest and promote mutual cooperation.[5]

At the closing session of the OPEC summit, on March 6, 1975, Bumidian announced the news of the agreement between the two leaders of Iran and Iraq, which he said had ended the conflict between "the two brotherly countries." The surprise announcement was received with jubilation, and the audience rose to give the two leaders, as well as the head of the host country, a standing ovation for the success achieved at Algiers. The OPEC summit thus could claim credit for the reconciliation between Iraq and Iran. "The final settlement of the thorny disagreement between Iraq and Iran," in the words of OPEC summit records, "which was achieved during the night of 5/6 March in Algiers, bears unsurpassably clear witness to the maturity, solidarity and potential of the Organization."[6]

On his return to Iran, the Shah claimed that he had fulfilled one of his longstanding national demands.[7] Saddam Husayn, no less satisfied with the outcome of the OPEC meeting, submitted the Algiers Agreement to a joint meeting of the Revolutionary Command Council (RCC) and the Regional Command of the Baʿth party on March 10, 1975, for a scrutiny of the text. It was ratified by all present in accordance with the Interim Constitution (Article 43) and authorized the country's foreign minister to undertake implementation of the agreement. Saʿdun Hamadi, Foreign Minister of Iraq, and ʿAbbas ʿAli Khalʿatbari, Foreign Minister of Iran, met in Tihran on March 15, and signed a protocol two days later that provided for the establishment of three committees: the first to examine the demarcation line of Shatt al-Arab; the second, the land boundaries between the two countries; and the third, the ways and means of preventing infiltrations across borders. The committees were to report to another meeting of the two foreign ministers within two months. Prime Minister Amir ʿAbbas Huwayda (Hoveida) of Iran, in a official visit to Baghdad from March 26 to March 29, carried on further talks, which confirmed the Algiers Agreement and laid down steps for its implementation.

In June 1975, the draft of a "reconciliation" treaty, designed to embody the principles laid down at Algiers to settle all outstanding differences between the two countries, was ready for signing by the foreign ministers of Iraq and Iran. Algerian Foreign Minister Butafliqa, who had played a constructive role in previous negotiations, joined the foreign ministers of Iraq and Iran in signing the treaty on June 13, 1975.

The treaty, made up of eight articles, embodied the general principles governing the settlement of the disputes between the two countries, which had been agreed on in the Algiers Agreement (Articles 1–3). It also stipulated that the provisions of this treaty and the protocols attached to it were "final and permanent provisions, irrevocable for whatever reason and . . . any encroachment upon any element of [it] . . . is contradictory in principle to the essence of the Algiers Agreement" (Article 4). The principle of nonencroachment was reiterated with regard to "the safety of national territories of both states" and, more specifically, with regard to the frontiers, which were considered "permanent and final" (Article 5). The final article dealt with the interpretation and implementation of the treaty and the protocols. Should any dispute arise between the two parties, it should be resolved in accordance with the following procedures:

1. By direct bilateral negotiations within two months from the time of application by either side.
2. By the good offices of a third friendly state, if no agreement was reached between the two parties within three months.
3. By arbitration, if either party declined to accept the good offices of another state within a month from the time of the rejection of good offices or failure to use it.
4. By arbitration tribunal, if arbitration were unacceptable within fifteen days from the time it was demanded by either party. The arbitration tribunal would be formed by the appointment of one judge by each part from one of its subjects, and the two judges would nominate an umpire. If the parties failed to agree on the appointment of the judges or the umpire, the chairman of the International Court of Justice would have the right to appoint the judges as well as the umpire within a period of one month from the time of their failure to act.

Four protocols were attached to the treaty. The first provided for the establishment of border security arrangements to prevent the infiltration of subversive and undesirable elements from one side of the border to the other (that is, all individuals regarded as *personae non gratae* by either regime). The protocol specified ways and means (including exchange of information) of how to cope with infiltration and subversion from either side.

The second protocol provided for the establishment of an Iraqi-Iranian-Algerian committee to carry out the redemarcation of the land border between Iraq and Iran, as it had been defined in (1) the Constantinople Protocol of 1913 and the minutes of the Turko-Persian Border Delimitation commission of 1914; (2) the Tihran Protocol of March 17, 1975; (3) minutes, dated of April 20, 1975, which approved, among other things, the minutes of the committee entrusted with the redemarcation of land borders in Tihran on March 30, 1975; (4) descriptive minutes, dated June 13, 1975, of the demarcation committee of the land border between the two countries; (5) maps, aerial photographs, and all other documents relating to the land frontier.

The third protocol stipulated that the Shatt al-Arab waterway border between the two countries be the thalweg rather than the eastern bank, as laid down in the Algiers Agreement of March 6, 1975, and confirmed in the Tihran Protocol of March 17, 1975, and the minutes of the April 20, 1975, meeting of the Foreign Ministers in Baghdad. Iraqi and Iranian vessels as well as vessels belonging to a third country would enjoy freedom of navigation, regardless of the thalweg, in all parts of the navigable channels of the Shatt al-Arab.[8]

The purpose of the treaty was thus not merely to settle border disputes, but also to put an end to longstanding issues between two neighbors whose mutual cooperation was necessary for internal unity as well as for regional peace and security. Retaliatory actions have often resulted in hurting the interests of the citizens of each side resident in the other, and each country had become involved in encouraging subversive elements against the other's regime. The Shah's support of the Kurds proved not only a danger to the Baʿth regime, but also a threat to national unity. If Mulla Mustafa had won the war, he probably would have pushed Kurdish claims far beyond autonomy. Had the Shah failed to come to an agreement with Iraq and put an end to the Kurdish War, Iraq would have been bound to seek Soviet support, which would have increased regional tensions and rivalry among the great powers in the Middle East. It was therefore an important landmark in Perso-Iraqi diplomatic relations, and will perhaps remain a significant document in any attempted future negotiations concerning the frontier between the two countries following the present war.

Two immediate steps were undertaken even before the treaty took effect. First, the Shah of Iran stopped his assistance to the Kurds and encouraged Mulla Mustafa to accept a cease-fire, announced by Iraq and Iran on March 13, 1975.[9] Iraq declared that amnesty would be granted to the Kurds, including deserters from the Iraqi army, and made it clear that the only way to "avoid further bloodshed" was to accept the terms of the general amnesty, which was to expire on April 1, 1975. Iran called on the Kurdish refugees who had been in Iran to decide on April 1 whether they wanted to remain in Iran or to return to Iraq, as the frontier between the two countries would be closed on that date. Only a few thousand returned by the end of March, but when the amnesty was extended to the end of April, a few thousand more returned, as both the Iranian and Turkish borders were closed to Kurdish infiltration. The Shah's prompt action to cooperate in putting an end to the Kurdish War greatly satisfied the Iraq government, and the Iraqi army took complete control of the Kurdish area by the end of March. Mulla Mustafa, like his ally the Shah, learned that he had to undergo medical treatment, and he surrendered to the Iranian authorities. Shortly afterward, he made his way to Washington, where he received medical treatment. He died of cancer in 1977.[10]

Second, the committees entrusted with the demarcation or the redemarcation of the land and water frontiers were set up to work in accordance with the second and third protocols shortly after the treaty came into force in the fall of 1975. While it was much easier to accomplish the work of the committee dealing with the water frontier (thalweg), to the great satisfaction of the Shah, the work of the committee dealing with the land frontier, presumably intending to return a number of sectors to Iraq, was considerably delayed by physical difficulties of the terrain and by bureaucratic red tape. By the time the Shah's regime collapsed, three sectors had yet to be surrendered by the Iranian authorities.[11] Moreover, Iran also began to claim certain privileges in the navigation of the Shatt al-Arab, which is the only outlet for Iraq to the Gulf, and sovereignty rights over waters for possible use for navigation by Iraq.[12] However, Iraq considered these matters to be of no great significance and to be resolved by negotiations, since the primary objectives were to establish the principle of good-neighborly relations and to put an end to infiltration and interference in domestic affairs.[13]

The Shah's preoccupation with domestic affairs and the bureaucratic reluctance to speed up the implementation of the treaty of 1975 inclined Iraq to take an initial favorably attitude toward the newly established regime in Iran by the Islamic Revolution. The Ba'th leaders seem to have believed that the new Iranian regime might cooperate with Iraq were they to recognize it and open a new chapter of trust and good-neighborly relations between the two countries. Iran's new leaders, however, would have nothing to do with the Ba'th regime. They not only ignored Iraq's initial gestures of friendship, but also began to make unfavorable statements about Iraq and its leaders, both in high official circles and in the press and media. Very soon, it became clear that the Iranian revolutionary regime was prepared neither to accept the treaty of 1975—much less to implement its provisions—nor to recognize the Ba'th government.

The new Iranian regime made no official decision concerning the treaty of 1975, but several men in high official circles—including the prime minister, foreign minister, and representatives to the United Nations—made statements in which they declared that the treaty with Iraq, signed by the deposed Shah, was not binding on Iran.[14] When these statements were reiterated in the press, the Iraqi foreign minister dispatched a letter to his Iranian counterpart in which he inquired whether the statements made by the

Iranian officials concerning the Treaty of 1975 represented the position of the new Iranian regime. Iraq's question was never answered.[15]

Was the treaty considered null and void, it may be asked, by the mere fact that Iran showed no inclination to honor the provisions relating to the sectors of the land frontiers that Iraq expected to be returned to it? Under International Law, a treaty is still valid and binding if it is denounced by one party without the approval of the other party or parties signatory to it. Moreover, the Treaty of 1975 stated that its provisions (including the protocols attached to it) "shall not be infringed under any circumstances," and "a breach of any of the components of this over-all settlement shall clearly be incompatible with the spirit of the Algiers Agreement" (Article 4). Since Iran failed to observe the provisions relating to the land frontier and refused to return the three sectors to Iraq, should Iraq surrender to Iran its sovereignty over half the waters of Shatt al-Arab? Iraq declared the treaty null and void as a formal step to relieve itself of its obligations before it sought to restore its jurisdiction over the three land sectors that Iran failed to surrender in accordance with the treaty.[16] Thus the spirit of Algiers, hailed by its signatories to have opened a new chapter in good-neighborly relations, vanished almost as soon as the ink with which it was signed had dried.

# VII

## Changes in the Regimes of Iran and Iraq and Political Developments Leading to the Gulf War

Hardly four years had passed since Iran and Iraq had reached a peace settlement in Algiers (1975) when significant changes occurred in the political structure of both countries, which had an adverse effect on their newly established friendly relations. In January 1979, the Shah's regime was overthrown by the Islamic Revolution; six months later, the presidency of Iraq passed from Ahmad Hasan al-Bakr to Saddam Husayn. It is not our purpose to inquire into the causes and the events leading up to the establishment of the Islamic Republic in Iran, or into the structure of the Iraqi political system, yet the changes in the structure of Iraqi politics following the transfer of the presidency from Ahmad Hasan al-Bakr to Saddam Husayn—events that are far less well known to Western readers than the establishment of the Islamic Republic in Iran—had far-reaching effects not only on Iraq's relations with Iran, but also on Iraq's relations with Syria and ultimately on the development of the Gulf War.

We have already observed that Persia, before it emerged as an independent state, was part of the Great Society of Islam, which regarded all countries of unbelievers as being in a "state of war" (to borrow a modern legal concept) with Islam. But since it seceded from Islamic unity and adopted Shi'ism as the official religion, its conduct of foreign relations was bound to change significantly in order to be accepted as a member of the modern community of nations. Today, by reviving classical Islamic standards, the Iranian Revolution has sought to reverse the development set in motion earlier when Persia seceded from Islamic unity and gradually began to adopt modern secular standards. No boundaries that separate one Muslim state from another, under the Islamic political system, were recognized. The Imam or his deputy (since the Imam's rule is in abeyance, according to the Shi'i creed) is, in theory, the supreme ruler over all believers, regardless of territorial segregation. All other rulers, civil or otherwise, are subordinate to the Imam's authority. Today, the Ayat-Allah Khumayni, considered the supreme religious authority in Iran, plays the role of Imam (or his deputy) and claims the right of calling on believers all over the world to rise up against foreign domination, oppression, and unjust rulers.

No essential differences exist between Sunni and Shi'i teachings with regard to the relationship between believers and unbelievers, according to Khumayni. His teachings seem to reinforce the teachings of spiritual leaders in other Islamic lands who have also called for the reestablishment of Islamic standards and have denounced oppression and foreign pressures. Thus it is not surprising that after the establishment of the Islamic Republic in Iran, Khumayni's call on scholars in other countries to stir similar upheavals alarmed Muslim rulers, especially extremist elements that had in recent years become very active.

After the Islamic Republic had been established in Iran, the Iranian leadership—perhaps not without Khumayni's tacit consent—began to reassert the principle of legitimacy of the Imamate (that the Imam must be a descendant of the Caliph 'Ali in direct line), which is unacceptable to Sunni followers, who insist on the principle of the election of the Imam (or Caliph) by the public. Since the last of the Shi'i Imams had disappeared in the ninth century, but will eventually return as the Mahdi (Messiah), it was taken for granted that the mujtahids would provide guidance for believers during the Imam's absence. In Iran, it had become the rule for the Shah to seek the advice and guidance of the mujtahids on matters of law and religion, although in practice he often disregarded advice. Not all mujtahids, however, stood together against the Shah when Western secular innovations, injustices, and other corrupt practices crept into the civil administration, since some held that their task was only to give advice, not to engage in political activities.

Khumayni, an activist, appealed to all mujtahids that their duty was not merely to advise; for if their advice were disregarded, they were duty-bound to assume authority and put an end to the Shah's oppression and misguided policies. In his lectures on Islamic government, given at Najaf before the revolution, he proposed to establish a regime presided over by the faqihs (jurists), called the wilayat-i faqih (governance of the jurists), and assume all powers in accordance with the Shari'a, which should supersede civil legislation and become the basis of all political decisions, including foreign affairs.[1] Following the Iranian Revolution, a Supreme Council, over which

Khumayni presided, was established, presumably to exercise the functions of the wilayat-i faqih and to guide the Islamic Republic, whose duty is to put into practice the higher authority of the Imam. Khumayni's immediate concern was, of course, with conditions in Iran, but his call for the establishment of an Islamic government was a message to scholars in all Islamic lands. His lectures on the subject were published while he was still in exile, and his ideas and ideals rose above national denominations, since he spoke with the voice of an Imam preceding the era of nationalism and his call was addressed not only to Shiʿi followers but to all believers.

Following the Iranian Revolution, Khumayni's ideas and messages were identified with Iranian Shiʿism, which is often associated with Persian nationalism. Indeed, the deeper he became involved in domestic affairs, the more clearly his activities seemed to reveal Persian traditions, especially after Iran had become involved in the war with Iraq; the new Iranian regime could no longer claim to fulfill the promises of the Islamic Revolution, which initially had received support from many elements outside clerical circles. Three fundamental factors may be said to have significantly affected the character of the Islamic Revolution and reduced it to the national level.

First, Khumayni's insistence that the supreme authority of the Imamate—the original cause of the split of Islam into two major confessional divisions—be exercised in accordance with the Shiʿi doctrine prompted Sunni scholars to disagree.[2] Before the revolution, Khumayni avoided entering into the polemical differences about the Imamate between Shiʿi and Sunni scholars. Since his return to Iran, however, Khumayni has often been addressed by his followers as the "Imam," presumably implying that he is either a "deputy" of the Imam in occultation or the head of the collectivity of mujtahids represented in the Supreme Council of the Islamic Republic. In either case, the appellation was acceptable neither to Sunni scholars nor to some of his Shiʿi peers, who rejected his interpretation of Shiʿi doctrines, especially the concept of wilayat-i faqih. Indeed, there are some who think that he is not qualified to be called even "a deputy Imam."

Second, Khumayni's initial call to believers in other Islamic lands to participate in the Islamic Revolution aroused the enthusiasm of extremists in Egypt, the Gulf region, Pakistan, and other Islamic lands—indeed, some even made an attempt to seize power by violence—but his involvement in the Iranian Revolution and the struggle for power among Iranian revolutionary leaders tarnished his spiritual leadership with narrow confessional, if not strictly national, bias. True, in order to ensure the success of the Islamic Revolution, he was bound to give support to Iranian revolutionary leaders in order that Iran would be the fountain spring for other revolutions elsewhere. However, instead of rising above rival groups, he sided with leaders who represented extreme clerical views. When the extremists won, the moderate leaders fled the country and warned their countrymen from exile about the dangers of clerical extremism. Perhaps no less important is that Khumayni, although thoroughly at home with Islamic learning, was out of touch with modernist movements. Unlike Jamal al-Din al-Afghani, who traveled widely in Islamic lands (as well as in Europe) and addressed himself to Sunni and Shiʿi followers alike, Khumayni's travels abroad were relatively limited—he spent hardly a year in Paris after leaving Najaf—and he returned to Iran in his old age with no intention of leaving it again. In both word and action, he gave the impression that his concern was primarily with traditional Islamic (Shiʿi) affairs, although this may not have been his original intent. Thus he lost the support of Islamic modernist elements.

Third, Khumayni's doctrine of the "export of the revolution," proclaimed after the Shah's regime collapsed, was perhaps at first intended to spread the revolutionary movement by peaceful means. In 1972, while he was still in exile in Najaf, he gave lectures on jihad—the Islamic concept of war—in which he stressed its defensive aspect and emphasized in particular the struggle of the individual against evil. Above all, he said, the ultimate aim is the purification of the soul.[3] This apsect of the jihad, in contrast with the view that jihad stresses resort to violence, is often called the "greater jihad" (al-jihad al-akbar). After the establishment of the Islamic republic, however, Khumayni made no secret of his intention to spread the Islamic Revolution to other countries by violence, which aroused the concern of his next-door neighbors. It is true that the duty of jihad, according to classical texts, might be fulfilled by peaceful as well as violent methods, and the defensive and offensive aspects of war became virtually indistinguishable, especially in wars against the territory of unbelievers.[4] But Khumayni's objective of "exporting the revolution," were it to be in fulfillment of the jihad duty, should be directed against unbelievers outside the house of Islam and not against believers inside, according to authoritative texts.[5] Khumayni's followers, not without his encouragement, seem to have been carried away by the celerity with which the Islamic republic was established and aspired to export its blessings to neighbors at the earliest possible moment. In their eyes, the regimes presided over by secular rulers, such as the Baʿth in Iraq and the dynastic regimes in the Gulf, had fallen under foreign influences and therefore, like the Iranian regime under the Shah, were unworthy of survival. They have accordingly been denounced and considered lacking in legitimacy and liable to be overthrown as a religious duty, presumably in accordancee with the Quranic injunction that states, "O believers, fight the unbelievers who are near to you, and let them find harshness in you" *(Qur'an IX, 125).*[6]

But the "unbelievers" in question, as noted earlier, were not the subjects of an Islamic state (unless they became renegades) but the unbelievers outside the territory of Islam.[7] For this reason, Iraq and other Gulf states considered Khumayni's doctrine of the export of the revolution to be irrelevant and a threat to their legitimate regimes, not to mention a violation of territorial sovereignty and an interference in the domestic affairs of their countries by stirring local Shiʿi elements against them.

At the outset, the news of the overthrow of the Shah's regime by the Islamic Revolution was received in Iraq and other Gulf states with mixed feelings. On the popular level, the reaction to the Shah's fall was favorable, since the clerics were at first allied with younger leaders who had received their education in Western institutions; it was hoped, however, that the new leaders would follow a foreign policy sympathetic to the aspirations of other Muslim countries as well as to the Arabs, as both Khumayni and the younger leaders made statements in support of the Palestine Liberation Organization (PLO) and other groups. Likewise, in high official circles of Iraq, some of the Baʿth leaders, critical of the Shah's hegemonic policy, held that the Iranian Revolution might provide an opportunity of trust and cooperation, since the new leaders denounced foreign alliances and wished to pursue a nonaligned foreign policy. From February 1979, when the revolution was still in its infancy, to July 1979, when the presidency of Iraq passed from Ahmad Hasan al-Bakr to Saddam Husayn, the question of whether Iraq should take a favorable or noncommittal attitude to Iran was often debated in closed meetings of the Revolutionary Command Council (RCC) and the Baʿth Regional command. Some of the Baʿth leaders argued that the new regime was not expected to alter Iran's hegemonic policy toward Iraq and the Gulf region and suggested taking a firm stand against it. Others held that both governments advocated several identical goals—full independence, nonaligned foreign policy, and other revolutionary principles—that might be the basis for cooperation between them. President Bakr is reputed to have taken the latter position; but others counseled patience, preferring to wait until the official attitude of the new regime toward Iraq had become clear.

Following the establishment of the Islamic Republic, the viewpoint of the moderates who sought to accommodate to Iran and enter a new era of normalization and cooperation prevailed. But the new Iranian regime seemed prepared neither to deal with Iraq on the basis of equality and reciprocity nor even to maintain the minimum normal relationship between two neighbors. Nevertheless, President Bakr, hoping that internal conditions in Iran might improve and that moderate leaders would assume control, counseled patience. But political trends in Iran were not very encouraging, and the extremist clerics seemed already on their way up the political ladder. Small wonder that several Baʿth leaders were alarmed that the passing of leadership into the hands of Iranian clerics might encourage opposition groups in Iraq to agitate against the Baʿth regime. They recognized the need for stronger hands to protect the regime from the infiltration of revolutionary elements into the country.

In these circumstances, the political trends in Iraq—indeed, in many other Arab countries—began to change, and relations between Iraq and Iran took a turn for the worse. Two crucial events led to the transfer of power from President Bakr to Saddam Husayn, and the advocates of a firm stand toward Iran came to the fore.

First, President Anwar al-Sadat's historic journey to Jerusalem in 1977, initially intended to resolve the Arab-Israeli conflict by peaceful methods, aroused the concern of other Arab countries that it might lead to a separate peace between Egypt and Israel. At a meeting of the Arab summit in Baghdad (November 2–5, 1978), they suspended Egypt from membership in the Arab League and broke off diplomatic relations with it.[8] Saddam Husayn, Vice President of the Iraq Revolutionary Command Council, played an important role at the Baghdad summit that brought him into the limelight in inter-Arab affairs. Moreover, Syria, an ally of Egypt in the October War against Israel (1973), took a firm stand against Sadat and, in cooperation with Iraq, persuaded other Arab countries to boycott Egypt following Sadat's negotiations with Israel. Amid those events, Syria and Iraq initiated negotiations at the highest levels for possible unity between the two countries to provide effective leadership in inter-Arab affairs. These negotiations, which continued until 1979, were suddenly interrupted by side issues, to which we shall return.

Second, while the unity negotiations were in progress, President Bakr, long in poor health, felt compelled to advise the RCC about his intention to relinquish the seals of office. Rumors were at once set afoot that he was under pressure to retire, since the events in Iran and the negotiations with Syria were not proceeding to the satisfaction of leaders at the higher echelons of the Baʿth party. Indeed, it was not the first time that Bakr wished to retire, as he suffered from a number of illnesses—including high blood pressure and diabetes—and wanted to attend to his health. Five years earlier, President Bakr was thinking about retirement for health reasons, but senior party leaders persuaded him to remain in office.[9] It is not surprising that as the pressure of work increased, Bakr felt that the country needed a more determined leadership and that the time had come to resign. It was taken for granted that Saddam Husayn would succeed him, and President Bakr himself expressed his wish to nominate Saddam Husayn to the presidency. But the rush of succession, carried out in the wake of the complicated negotiations for unity with Syria, gave rise to speculations that perhaps it had been forced on Bakr. Before we discuss the transfer of power from Bakr to Saddam Husayn, perhaps a brief account of the unity negotiations might be useful.

On October 1, 1978, in preparation for the Arab summit meetings in Baghdad, the RCC passed a resolution in which it was decided (1) to call an Arab summit, to be held in Baghdad, in order to discuss possible steps to be taken to counteract the impact of the Camp David Agreement; (2) to call on Syria and Iraq to cooperate against possible confrontation with Israel; (3) to call on all Arab countries to form a front against the implementation of the Camp David Agreement and to provide adequate funds for a period of ten years.

No sooner had the RCC resolution been proclaimed than President Hafiz al-Asad of Syria sent a moving letter to President Bakr in which he expressed his concern about the lack of cooperation among the Arabs in the face of an impending peace plan for the Arab-Israeli conflict. He went on to argue that at no other time had the need for unity between Iraq and Syria been more urgent; they had to stand together against Sadat's move to make a separate peace with Israel. If Egypt were capable of making peace with Israel, Asad asked, how could Iraq and Syria—two sister Arab countries—fail to cooperate against their common enemy while they claimed to belong to one party (the Ba'th) and advocate the same national goals? Bakr, it is reported, was so touched by Asad's appeal that he at once ordered the initiation of negotiations for unity between the two countries. Saddam Husayn, then in Saudi Arabia attending an Arab summit meeting, returned just in time to participate in a meeting of the RCC, the Regional and National Commands, in which Asad's letter was discussed. It was unanimously agreed to reopen negotiations with Syria—this was not the first time Iraq and Syria had discussed unity—and to discuss the steps to be taken to bring the two countries together under some form of union. Tariq 'Aziz, Foreign Minister of Iraq, was charged with the task of sounding out President Asad's views about the form of union he wished to be established. 'Aziz left Baghdad for Damascus on October 5, 1978, and met with President Asad and other Syrian leaders on the same day. Having been informed about the action taken in Baghdad, Asad was asked by 'Aziz, "What kind of union would you propose?" Asad, begging the question, replied that he wanted Iraq to submit the unity proposals. "Give my greetings to our comrades," he added, "and tell them that we are prepared to accept any form of union, ranging from saying 'welcome' *[marhaba]* to 'full unity.' "

On his return to Baghdad on the same day, 'Aziz reported Asad's reactions. The Iraqi leaders decided at once to prepare a draft "pact of national action'' to be presented to the Syrian leaders. In that draft, it was proposed to establish the Higher Committee, composed of fourteen members—seven members from each country—which would ultimately decide the form of unity agreeable to both countries. Moreover, the Iraqi leaders suggested the establishment of several committees to consider all matters of mutual interest, to be submitted to the Higher Committee. Five committees were proposed: (1) Committee for Political Affairs; (2) Committee for Cultural and Informational Affairs; (3) Committee for Economic Affairs; (4) Committee for Educational Affairs; and (5) Committee for Military Affairs. Armed with the draft National Pact and a covering letter from President Bakr, ʿAziz left Baghdad for Damascus almost a week from his first visit. He remained overnight, waiting for the reactions of the Syrian command. On the following day, he was told that the Iraqi proposals were agreeable to the Syrian command in principle. Informed that very soon Asad himself would visit Baghdad for further talks on the Iraqi proposals, ʿAziz left the Syrian capital at once.

On October 24, 1978, President Asad, at the head of a large delegation, arrived at Baghdad. during the following two days, several meetings were held in which the Iraqi proposals were discussed. A revised version, entitled "Declaration of a National Pact,'' was made public; both Iraq and Syria declared themselves ready to enter into unity, the form of which, whether federal or unitary, would be subject to further scrutiny, as certain matters, such as the correlation of the two branches of the Baʿth party, were in need of further clarification.[10]

On November 2, 1978, when the Arab summit was holding its meetings in Baghdad (to discuss measures to be taken against Sadat for making peace with Israel), a misunderstanding about the meaning of the word *one* developed between the Syrian and the Iraqi leaders when President Bakr, in a visit to the Syrian delegation, told Asad, ''We [Syrians and Iraqis] should deal with other Arab delegations as if we are one state.'' Bakr's statement, taken literally by the Syrian delegation to mean that the Iraqi leaders were in favor of a unitary state, was used in a figurative sense to imply that the two countries would be in one camp in their relations with the other delegations. This incident, breeding suspicion between the two delegations, prompted the Committee for Political Affairs to resume talks on the form of unity agreeable to both countries.

Since the Committee for Political Affairs met first in Baghdad, its second meeting was held in Damascus in January 1979. Saddam Husayn led the Iraqi delegation. At the committee's first meeting, the form of unity was the first item on the agenda, as Saddam Husayn was anxious to know what kind of unity Asad had in mind. But Asad was not forthcoming: he simply remarked that he would like "A unity in which neither side would be the loser." Thereupon, Tariq ʿAziz and ʿAbd al-Halim Khaddam, Foreign Ministers of Iraq and Syria, met to discuss the formula that would be agreeable to both sides. It was proposed that the minimum level of unity should be a form of "federal union." The Iraq delegation was entrusted with the task of preparing the proposals for the unity of the state, and the Syrian delegation the proposals for the unity of the Baʿth party.

The Iraqi delegation met at once and drew up a draft plan for the establishment of a federal state called the "Federal Arab State," consisting of two "regions." Each region, it was proposed, should have its own government, composed of a regional president, a council of ministers, a regional command council, and a legislative assembly. The Syrian delegation, which was expected to prepare a specific plan for the unity of the Baʿth party, submitted instead three draft plans for the unity of the state but none for the unity of the party. The three proposals were a confederative plan, like the plan of the then existing unity among Egypt, Syria, and Libya; a unitary state, like the defunct United Arab Republic; and a loose federal state. Disappointed by the vagueness of the Syrian plans, the Iraqi delegation inquired why the Syrian delegation had not submitted a specific plan for the unity of the Baʿth party. In reply, the Syrians pleaded that the matter was very ticklish, in view of the fact that there were a few Syrians in the Iraqi Baʿth party, such as Michel Afalq and others. "But there were also a few Iraqis in the Syrian Baʿth party," quipped one member of the Iraqi delegation. Nor did the Syrian committees, save the Committee for Educational Affairs, prepare any proposals for coordination on matters of mutual interest. Thus the meetings of the Higher Committee in Demascus proved very disappointing, and the Iraqis returned to Baghdad before the preparatory work on unity was completed.

On June 24, 1979, President Asad, leading a Syrian delegation, suddenly arrived in Baghdad. He announced that the Syrian delegation was ready to continue discussions of both the unity of the two countries and the unity of the two branches of the Baʿth party. At first, a closed meeting between Asad, representing Syria, and Bakr and Saddam Husayn, representing Iraq, was held. After a protracted discussion, the three leaders invited Foreign Ministers Tariq ʿAziz and ʿAbd al-Halim Khaddam to join them and told them that they had agreed on the formula "the United Arab Republic" as the form of unity between the two countries, but they had not yet agreed on the form of the presidential council. After further discussion, Asad proposed a council composed of two members: one Syrian and one Iraqi. Bakr and Saddam Husayn proposed three members: one Syrian and two Iraqis. As a compromise, Tariq ʿAziz proposed a council of seven: four Iraqis and three Syrians. But no final agreement was reached.

At a meeting attended by all members of the Syrian and the Iraqi delegations, the federal parliament and the federal budget were discussed. The Syrians proposed that the whole income from oil be included in the federal budget. The Iraqis, realizing that almost all the oil income would be from Iraq, proposed that the oil income from each region be left to the regional budget, and suggested that only a certain percentage of the oil income from each region (to be agreed on later) be earmarked for the federal budget. With regard to the composition of parliament, the Syrians proposed that membership be divided equally between the two countries, but the Iraqis insisted that parliament should not represent regions, but the population as a whole. While there was some flexibility on other matters, the two sides could not agree on the composition of parliament. However, before the Syrian delegation left Baghdad, the Iraqi delegation submitted a draft plan of unity composed of the following: a president of the federal state, a vice president, a command council (composed of fourteen members equally divided between the two countries), a federal government (composed of seven departments: Foreign Affairs, Defence, Information and Culture, Planning, Education, Justice, and Finance), a parliament (composed of members elected on the numerical strength of the population in each country), a regional government in each country, and a federal constitution. The question of the unity of the Baʿth party was left open, either to be united into one or to remain divided into two branches, one in each region.

The rock on which the whole unity plan was wrecked was the question of the presidency, which, although often discussed among the members of each delegation, was never officially put on the agenda for discussion. Nevertheless, each delegation was aware of the views of the other. The Iraqis in particular insisted that the president be from Iraq because of the country's potential in natural and human resources and its impressive historical and cultural role in Arab traditions. These views were made crystal clear in unofficial contacts, although the Syrian leaders, considering the Iraqi views as an expression of narrow local traditions, never really expressed their own opinion on the matter. Differences on the presidency and the composition of parliament irked the Syrian leaders and often proved to be the cause of disagreement on less important matters, such as regional frontiers and the choice of the federal capital. While negotiations were still in a deadlock, Asad suddenly left Baghdad on the pretext of the sudden Muslim Brothers' attack on the Aleppo barracks, which stirred confessional feeling in the country, but his real reason for leaving was the differences over leadership.

The unity negotiations were never resumed.[11] Indeed, relations between Syria and Iraq began to deteriorate and often led the two countries to resort to violence, especially after the Iraqi presidency passed from Bakr to Saddam Husayn. Above all, the Syrian leadership was accused of interfering in the transfer of power from Bakr to Saddam Husayn. When Iraq went to war with Iran, the Syrians took the side of Iran, largely because of rivalry on leadership.

Hardly a month after Asad's departure from Baghdad, President Bakr suddenly made up his mind to retire. Although it was rumored that he may have been pressured to retire because of differences of opinion on certain matters of foreign policy, Bakr was indeed in poor health and insisted on leaving office. In order to avoid giving the impression that there were personal differences among high-ranking leaders, Bakr himself suggested that he nominate Husayn as his successor.

There was, however, a certain constitutional ambiguity about the election of the President of the Republic. Article 38 of the Interim Constitution provided the method of electing the president as follows:

> The Revolutionary Command Council . . . by a majority of two thirds . . .
> (a) elect the chairman from among its members who shall be designated as the Chairman of the Revolutionary Command Council and President of the Republic.

This clause makes no mention of the president's term in office or of the possibility of his resignation. It seems that it was taken for granted that the president would serve for life and that the RCC would elect by a two-thirds majority any one of its members as president. Should the president resign, however, could he nominate the vice president for possible election by the RCC? It is possible that these and other matters might have crossed the minds of some members of the RCC, before it met to discuss the resignation of President Bakr. At the outset, Saddam Husayn, owing to his high personal qualities and qualifications, was the natural candidate for the presidency. Nor was there any other candidate who might be nominated as a rival to Saddam Husayn. Several informal talks among members of the RCC about succession must have taken place immediately after President Bakr expressed a desire to resign, but outwardly there seems to have been no differences of opinion among them.

On July 11, 1979, President Bakr called a meeting of the RCC, over which he presided, and announced in no uncertain terms that he wished to relinquish the seals of office on the grounds of poor health and inability to perform his duties adequately; he then submitted his resignation from all official functions. Following his brief presentation, Bakr withdrew, and Saddam Husayn, as vice president of the RCC, chaired the meeting. Although Bakr's wish to retire was not unknown to other leaders, they all regretted his action, but respected his wish to resign.

On the following day, the RCC met again to elect a new president, in accordance with the terms of the Interim Constitution. Saddam Husayn was nominated and elected unanimously as President of the Republic and Chairman of the Regional Command Council of the Ba'th party. 'Izzat Ibrahim al-Duri, a senior member of the Ba'th party, was elected vice president. It was decided that the elections of the president and vice president would be made public on July 16, and that Bakr would announce his resignation and the election of Saddam Husayn in a speech to the nation on that day. On July 17, celebrations and a procession would be held to commemorate both the anniversary of the July Revolution and the elevation of Saddam Husayn to the presidency.

No sooner had these arrangements been made than Muhyi al-Din 'Abd al-Husayn Mashhadi (known to his peers as al-Shammari), chief of the Presidential Office and Secretary General of the RCC, unexpectedly voiced his misgivings about the elections: "Why was the resignation of President Bakr so quickly accepted?" The implication was that Bakr should have been persuaded to stay, as, indeed, he had previously agreed to stay when requested to do so. This seemingly casual remark was taken rather seriously by most members of the RCC because it was made after the elections of Saddam Husayn as president and 'Izzat al-Duri as vice president were over. It signaled that Muhyi al-Din was dissatisfied with the elections—indeed, it virtually meant rejection of the actions that had just been taken by the RCC—in which he had participated. Had he asked his question on July 11 at the meeting of the RCC when Bakr submitted his resignation, his dissent would have been in order and might have been discussed in the spirit in which it was presented, although it is doubtful that the majority would have changed its mind. But objecting to the acceptance of Bakr's resignation during the meeting of the RCC after the president and vice president had been elected raised suspicion that his second thoughts might have been the product of a plot, since he had been seen during the day between the two meetings of the RCC speaking privately with Muhammad 'Ayish, former Minister of Industry, and perhaps other members of the Ba'th party. He was arrested at once after the meeting of the RCC on July 12. The RCC, resuming its meeting on the same day, decided to expel Muhyi al-Din from membership in the RCC and turned him over to the Special Investigation Committee, appointed by the RCC to investigate his dubious activities. The RCC then completed the formation of a new cabinet. In addition to the presidency, Saddam Husayn was entrusted with the post of prime minister. He was to be assisted by three deputy premiers, appointed from among members of the new cabinet.

On July 16, as had been arranged, Hasan al-Bakr made a speech to the nation in which he announced his resignation from the presidency and the election of Saddam Husayn as the new president. He paid high tribute to his successor as the person most suitable for the presidency.[12] The news about the formation of a new government and the names of its members were also broadcast to the nation. Meanwhile, ʿAdnan al-Hamdani, one of the newly appointed deputy premiers and Minister of Planning, was already on his way to Damascus to convey officially to the Syrian leadership the news about Saddam Husayn's election to the presidency.

On July 18, Saddam Husayn called to his office some of the senior members of the RCC to disclose the findings of the Special Investigation Committee about Muhyi al-Din Mashhadi's plot. Mashhadi had admitted to the committee that the remark he had made about the resignation of Bakr reflected not merely his personal opinion, but also the opinion of four other members of the Baʿth party: ʿAdnan al-Hamdani, the new deputy premier and Minister of Planning; Ghanim ʿAbd al-Jalil, new Chief of the Presidential Office and former Minister of Higher Education; Muhammad Mahjub, Minister of Education; and Muhammad ʿAyish, former Minister of Industry. Mashhadi confessed that he and his four associates had secretly come to an understanding that they should take control of the regime with the assistance of the Syrian government, which promised to extend both financial and military support. ʿAyish, the most assiduous of the four, acted as liaison between his group and the Syrian leadership.

At a meeting of the RCC, Saddam Husayn brought Mashhadi to testify before his former peers. He reported that Mashhadi had told the Special Investigation Committee that his four associates held views similar to his about the transfer of authority from Bakr to Husayn. Since ʿAyish was the most active of the group, he was also arrested and brought to testify before the RCC. Like Mashhadi, he admitted his role in the conspiracy. The RCC held another meeting, to which ʿAdnan al-Hamdani and Ghanim ʿAbd al-Jalil were invited. They were interrogated, but denied any complicity in the conspiracy. They were turned over to the special committee for investigation. The five plotters then were brought to another meeting of the RCC; Mashhadi and ʿAyish pleaded guilty and implicated participation of the others in the plot, but Hamdani and ʿAbd al-Jalil denied any complicity.

Within a week, the Special Investigation Committee completed its investigation, and a report confirming the alleged charges against the five plotters was submitted to the RCC. A special court, under the presidency of Naʿim Haddad, a member of the RCC, was appointed, and the five accused plotters were turned over for trial in accordance with the Interim Constitution (Article 387). The court found them guilty of conspiracy against the newly elected president. They and a few others who had been directly involved in the plot, including ʿAbd al-Khaliq al-Samarra'i (who was in prison for involvement in an earlier plot),[13] were condemned to death; the executions were carried out soon after the sentencing. Other plotters received various prison sentences, but the rest were acquitted. In all, twenty-two people had been involved in the conspiracy.

The incident brought to the surface the struggle for power in higher circles of the Baʿth party. Two possible interpretations may be offered to explain the plot; each stresses the viewpoint of one of the two groups in their struggle for power. One group represents local (Iraqi) interests. The other, reflecting an ideological conflict, may be called the Pan-Arab group. From the viewpoint of the latter, the five plotters were in favor of the presidency of Bakr because he displayed an understanding of and perhaps sympathy with the Syrian leadership to achieve unity; Saddam Husayn, however, seemed to lay greater emphasis on Iraq's role in order to assert his own leadership over the country. The other group, composed of the protégés of Saddam Husayn, saw in the continuation of Bakr's benign presidency a weakening of the Baʿth regime in the face of the growing danger from Iraq's next-door neighbor, although it had the highest respect for Bakr's personality and character. It was also suspected that the five plotters might have contemplated that were unity with Syria to be achieved under Bakr, President Asad, rather than Saddam Husayn would have the opportunity to succeed him as president of the union. Since three of the five plotters—Mashhadi, Hamdani, and ʿAbd al-Jalil—were Shiʿi followers, they were suspected to be in favor of a Shiʿi president, such as Asad, in order to enhance the position of the Shiʿi community in the union.

But this was not all. The majority in the RCC, meeting in a joint session with members of the Regional Command of the Baʿth party, was in favor of a strong presidency of Saddam Husayn. Only on July 12, the day after the election, did Mashhadi voice his dissatisfaction with the outcome of the election. Intelligence reached Saddam Husayn that when ʿAdnan al-Hamdani, who owed his elevation to a high position in the regime to Saddam Husayn, was in Damascus to convey the news of the transfer of the presidency, he had fallen under the influence of the Syrian leadership. What Hamdani really planned to do after his return to Baghdad will perhaps never be known, but the fact that he was a protégé of Saddam Husayn and he had just been reappointed as a deputy premier and Minister of Planning in the new government, must have been a great disappointment to the new president.

Finally, Mashhadi admitted at a meeting of the RCC that he and his associates agreed on their opposition to Saddam Husayn's candidacy to the presidency, although his confession has been construed as a promise that he would be spared capital punishment and perhaps ultimately be forgiven (probably after a short term in prison). But his admission has confirmed by ʿAyish, who also pleaded guilty before the RCC and the Special Investigation Committee. However, on July 18 and 19, when ʿAbd al-Halim Khaddam, Foreign Minister of Syria, was visiting Baghdad to convey the congratulations of the Syrian leadership to Saddam Husayn on his ascendancy to the presidency, he was informed about the alleged complicity of the Syrian leaders in the plot; but he denied Syrian involvement, despite being shown a video recording of Mashhadi's admission. In Damascus, the Syrian government officially denied any involvement and considered the whole plot a purely internal Iraqi affair.[14]

In Iraq, it became clear that if the regime were ever to survive under Baʿthist leadership, the party as a whole would have to rally behind Saddam Husayn, who proved the most effective leader since the party had come to power in 1968. As I noted in another work almost a decade ago, Saddam Husayn, the "heir apparent," has become "his Party's best promise for the country's future leadership."[15]

# VIII

# The Iraq–Iran War: The
# Diplomatic and Legal Aspects

The Treaty of 1975, designed to resolve pending issues and put an end to rivalry and conflict between Iraq and Iran, became itself the cause of tensions and the revival of conflict when the regime over which the Shah presided was overthrown in 1979. True, there were deeper causes of conflict between the two countries, such as the historic confessional tension and Kurdish demands for autonomy, which the treaty could not possibly have settled; but it certainly laid the ground for reconciliation and cooperation on the official level, which might have helped resolve other issues. The treaty, however, proved a great disappointment to Ayat-Allah Khumayni, because the Baʿth government not only stopped its support to the Iranian mujtahids in Iraq, but also imposed restrictions on their activities. Perhaps even more disappointing was the pressure brought to bear on Khumayni to discontinue his activities against the Shah, since Iraq had made peace with him. When he failed to heed Iraq's warnings, he was finally told in a cryptic message: "Shut-up or leave."[1]

From 1975, when Iraq signed the treaty with the Shah, to 1978, when Khumayni was expelled from Iraq, the Baʿth government virtually became the Shah's ally. But to Khumayni and his followers, the Shah was still the enemy, denounced as a corrupt, antireligious, and despotic ruler. They, therefore, never ceased to agitate against him until he and his regime had eventually been overthrown. Even before Iraq had signed an agreement with the Shah, Khumayni's fraternization with the Baʿth government was not intimate or intended to be far-reaching; it was a marriage of convenience, as both he and the Baʿth knew quite well that their cooperation served only immediate purposes and was not an agreement on fundamentals. Khumayni's lectures on the principles of Islamic government that he had given at Najaf were a far cry from the Baʿth ideology.

Following his expulsion from Iraq, Khumayni went to Paris, where he

79

enjoyed greater freedom of political expression and was joined by a host of
Iranian revolutionary leaders. The Shah's regime, exposed to a merciless
worldwide propaganda campaign conducted in the press and other media,
was rapidly disintegrating under shifting internal conditions. In February
1979, when the Shah left Iran, Khumayni returned triumphantly to declare
the establishment of the Islamic Republic, which he had envisioned while in
exile. A provisional government headed by Mahdi Bazargan, one of Khumayni's civilian followers, was set up to prepare the constitutional framework for the new regime. The mujtahids, as we have noted, moved slowly
but surely to dominate the political scene and finally brought the military
and civil authorities under their control. Khumayni's spiritual authority, exercised not as head of state but as spiritual leader and Chief of the Supreme
Council, prevailed. In other words, he became the real power in the land.
Although often called "Imam" by courtesy, he was not the historical Imam
of the Shiʿi community; he exercised his authority only as a "deputy," or
agent, of the Imam who went into the ghayba (absence), according to the
best commentators on Shiʿi doctrines.

The celerity with which the Shah's regime collapsed must have come to
the Iraqi leaders as a great surprise; indeed, it had come to many other
leaders elsewhere as no less a surprise. The question of the attitude that the
Iraqi government should take toward the new Iranian regime was often discussed in high political circles of the Baʿth party, but no quick decision was
taken. Two schools of thought may be said to have developed. One, advocating a confrontational stand, argued in favor of quick action; the other,
hesitating to act, counseled a nonentangling attitude. Most of the Baʿth leaders seem to have supported the latter school, on the grounds that the Iranian
Revolution was a purely internal affair with which Iraq had nothing to do.
Since the Shah had raised many difficulties for Iraq in the past, this school
maintained, it was not in Iraq's best interest to condemn the new regime
before its attitude had become clear. The advocates of the confrontational
school thus were persuaded to wait before taking action. But before the Iraqi
government should recognize the new Iranian regime, it was held, Iraq's
own policy toward its neighbor must be made clear in an official commu-
nication, although the very fact of sending such a document implied Iraq's
readiness to recognize Iran's new government.

On February 13, 1979, a memorandum addressed to the head of Iran's
provisional government was delivered by the chargé of the Iraqi embassy in
Tihran. The note, in which Iraq's policy toward the new regime was set
forth, may be summarized as follows:

1. Iraq's traditional policy, confirmed under the Baʿth government, has always been to pursue friendly relations with neighbors based on the prin-

ciples of respect for sovereignty, noninterference in domestic affairs, and the right of nations to achieve their legitimate national aspirations.

2. Iraq has always looked to Iran and Turkey with high regard in view of their Islamic and historic relationships with the Arabs in general and with Iraq in particular.

3. Iraq, which has experienced foreign domination and oppression, sympathizes with Iran and hopes that it will achieve its goals of liberty, justice, and progress.

4. Iraq derives satisfaction from the recent pronouncements of Ayat-Allah Khumayni and other Iranian leaders about their good intentions toward the Arabs and their denunciations of Zionism.

5. Iraq looks forward to cooperating with the new Iranian regime and hopes to deepen its friendly relations, promote mutual interests, and maintain stability and peace in the region, based on freedom and justice.[2]

On April 15, 1979, Ahmad Hasan al-Bakr, President of Iraq, sent a cable in which he congratulated Ayat-Allah Khumayni on the occasion of the proclamation of the Islamic Republic and reiterated Iraq's offer of friendly relations between the two countries. Bakr instructed the Iraqi ambassador in Tihran to call on Abu al-Hasan Bani-Sadr, President of Iran, to convey his congratulations. Later, Iraq extended an invitation to Premier Bazargan, head of the provisional government, to visit Iraq and enter into negotiations for normalizing the relations between the two countries. Iraq never received a reply to any of these communications, save a cable from Khumayni, communicated through the Iranian Foreign Office, and an edited text of the same cable, as it appeared in the Iranian press. The cable, though brief, was polite; but the edited text, different in tone, was discourteous.[3] Iraq twice attempted, without success, by means of direct contacts to initiate negotiations—the first by President Bakr, and the other by Foreign Minister Saʿdun Hamadi, who invited Premier Bazargan and Foreign Minister Yazdi to visit Iraq during the meetings of the nonaligned conference in Havana in 1979. While both Bazargan and Yazdi seem to have been prepared to deal with the Baʿth government, Ayat-Allah Khumayni warned his ministers that he was not in favor of negotiating with it. Nor was this all. Unofficial contacts were also attempted to bring about an understanding between the two countries, but nothing came of these personal efforts.[4]

Failure to bring the two neighbors to the negotiating table encouraged extremists in both to come to the fore. We have already noted how the moderates in Iran were replaced by extremists, who were at the outset determined to carry out Khumayni's doctrine of exporting the revolution. Iraq, a next-door neighbor, was considered the first target before reaching out to other lands. We have also seen how President Bakr, partly because of poor health, but also as a reaction to the trends of extremism in Iran, stepped down in favor of Saddam Husayn. Although Husayn was at first prepared to negotiate with Iran, he became quite concerned about the intentions of extremists who had been making declarations threatening Arab interests in the Gulf region. The need for a firm stand against the Iranian Revolution had already been deeply felt in the hierarchy of the Ba'th party, and the rise of Saddam Husayn to the presidency signaled that those in favor of a firm stand had won over those reluctant to make up their minds. Three stages in the strained relationships between the two countries may be discerned before Iraq went to war with Iran on September 22, 1980.

First, Iran's refusal to negotiate with Iraq aroused the concern of Iraq's leaders that the new Iranian regime was not prepared to honor the Algiers Agreement, which Iran had concluded with Iraq in 1975. Appreciating the initial preoccupation of Iranian leaders with domestic affairs, the Ba'th government was prepared to give the Iranian leaders the benefit of the doubt and began to dispatch official communications offering Iraq's good will toward the new regime and its readiness to open a new chapter of cooperation and friendship. Since no encouraging reply was ever received, the Ba'th government came to the conclusion that the Iranian leaders were prepared neither to honor their country's treaty obligations nor to deal with Iraq in accordance with normal diplomatic procedures.

Second, high-ranking Iranian leaders—including Ayat-Allah Khumayni, Sadiq Qutb-Zada, Muhammad 'Ali Raja'i, and 'Ali Akbar Rafsanjani—made statements to foreign correspondents in which the Ba'th leaders were accused of hostility toward Iran and denounced as oppressive, antireligious, and corrupt. Other declarations, revealing territorial ambition, stated that "between Muslim countries there are no borders" and that "all countries around the Gulf were historically part of Iranian territory." Above all, Khumayni's doctrine of exporting the revolution was often reiterated by Iranian leaders as the policy of the new regime. This doctrine, proclaimed on March 21, 1980, was at first embodied in a statement delivered by Khumayni's son on behalf of his father, in which it was stated, "We must . . . export our Revolution to other parts of the world and renounce the concept of keeping the Revolution within our own boundaries."[5]

The press and other media expressed public approval of Khumayni's doctrine in highly rhetorical style and presented it to to Iran's neighbors as an Islamic demand. But Iran's neighbors denounced it as inconsistent with Islamic teachings and stamped it as Shu'ubiya, the expression of Persian ethnicism. Khumayni's doctrine of exporting the revolution and his interpretation of other Islamic doctrines, often reiterated in the press, escalated the war of words that had already been taking place when the two countries went to war. We shall return to this subject later.

Third, the last stage in these developments was Iran's resort to violence. It began as incidents and skirmishes on the frontier before the armies of the two countries met on the battlefield. It is not our purpose to discuss the military aspect of war that is still in progress; only the diplomatic and legal aspects will be the subject of our inquiry, although military events that have a bearing on the subject will be considered.

Violations of air space were often reported in the press as early as February 1979, when the Iranian Revolution had just started.[6] Surely, these violations recurred on both sides of the frontier, but Iraq was not the first to resort to the use of force. Iran first launched a series of assaults on a number of cities within Iraqi territory, causing destruction of life and property, presumably intended to undermine the regime by arousing fear and frustration among the people. Perhaps the first important incident to occur in an urban center was the sudden explosion of a bomb at a peaceful gathering of students at the University of Mustansiriya in Baghdad on April 1, 1980. Very soon, this was followed by attempts on the lives of Tariq ʿAziz, Deputy Prime Minister, and Latif Nasif Jasim, Minister of Culture and Information. On April 5, another bomb, suspected to have been thrown from an Iranian school in the Waziriya Quarter, exploded in the midst of a funeral procession held in memory of those killed at the University of Mustansiriya four days earlier. On investigation, the Iraqi government uncovered a net of clandestine activists directed by Iran in various parts of the country. Infiltration and subversive activities were not new, but the government was alerted by those events and began to keep an eye on suspected persons, who were later brought to trial.[7]

During the summer of 1980, further sporadic violations of air space and incursions across frontier sections alerted the Iraqi authorities that Iranian forces had become very active. On September 4, the Iranian forces in the three central frontier sections—Zayn al-Qaws, Maymak, and Sayf Saʿd—which should have been returned to Iraq in accordance with the Algiers Agreement, began to shell several villages and towns, including Khanaqin, Mandali, Naft-i Khana within Iraqi territory. These hostile actions prompted the Iraqi government to dispatch an official note of protest to Iran through the Iranian embassy in Baghdad on September 7, and to warn that those unwarranted acts might lead to retaliation, but its warning was not heeded. Thus the Iraqi forces took control of Zayn al-Qaws to prevent violent attacks. On September 8, one day after the capture of Zayn al-Qaws, Iraq sent another note to Iran, demanding that it "give back the rest of the territories which Iran has encroached upon during previous periods . . . [so that] the two countries would avoid the possibilities of wider confrontations."[8] After Iran failed to heed these protests and warnings, the Iraqi government ordered its forces to enter the other sectors—Maymak and Sayf Saʿd—on the grounds that they were acknowledged as Iraqi territory in the Algiers Agreement and the Treaty of 1975. On September 11, Iraq dispatched a polite but firm note to Iran, which may be summarized as follows.

First, Iraq was concerned that the Iranian leaders, because of their preoccupation with domestic affairs, might "not be aware" of Iran's recurring "encroachments on Iraqi territory." These acts were contrary to both the Algiers Agreement and International Law. Iraq wished to call the attention of Iran's authorities to these incidents.

Second, the Iranian leaders, the note emphasized, should realize that assaults on Iraqi towns and villages by the armed forces of Iran "is a serious matter" and not "games" in which the Iranian military may be engaged without regard to Iraqi security. If these acts were not stopped, the note warned, the relationship between the two countries might deteriorate to a very dangerous situation.

Third, Iraq had no intention of attacking Iran, but if these incidents were not stopped, Iraq would be bound to use force as a means of defence in the areas under attack.[9]

Meanwhile, reports were reaching Iraq that the Iranian regime was in disarray and that the army and the people in the provinces were reluctant to support the leaders of the Islamic Republic. Those reports were provided mainly by Iranian fugitives (former officers and politicians) in Paris who maintained that the army was no longer in experienced hands, as the leading officers had been purged, and could not possibly win a war with Iraq. Two of them, Generals Shahpur Bakhtiyar and Ghulam ʿali Uwaysi (Oveisi), who had held high military ranks under the Shah, made several visits to Baghdad for consultation. They seem to have painted a gloomy picture about the army and conditions in the country. Were Iran to become involved in a war, they held, the new regime would face serious uprisings in the provinces. Like other exiled officers, however, they were out of touch with the country, and seem to have forgot the lesson of history that under the threat of war, no people would fail to respond to the call of duty to defend the homeland.

But Saddam Husayn, well known for his resolve and bravery, perhaps needed no great inducement to put an end to recurrent incursions across and violations of Iraq's borders. His assumption of power after Bakr's resignation a year earlier had inspired his regime with vigor and renewed determination to take a firm stand against a hostile neighbor. His ability and strength of character had already been acknowledged in the country. He was thus not expected to remain idle in the face of Iran's continuing encroachments on Iraq's security and territorial integrity. To Saddam Husayn, Iran's hostility was a challenge to his leadership. The response to this challenge was often referred to in the press as Saddam's Qadisiya—in memory of the decisive battle that destroyed the Persian Empire in 635 by the Arab army—in which Iraq has been engaged with Iran since early September 1980.

The dispatch of Iranian reinforcements to Zayn al-Qaws seems to have angered Saddam Husayn, since it was he who had reasserted Iraq's sovereignty over this and other central boundary sectors in his negotiations with the Shah at Algiers in 1975. This was an affront to Saddam Husayn; it was taken as a deliberate act to humiliate him and undermine his regime. He referred to it later in some of his public statements, and his foreign minister, Dr. Sa'dun Hamadi, called it in his speech to the United Nations General Assembly (October 3, 1980) the "turning point" in the chain of Iranian incursions across Iraq's borders.[10] But to Saddam Husayn, it was *casus belli,* as Iraq had attempted, in vain, to settle its differences with Iran by means of direct negotiations.

On September 17, 1980, Saddam Husayn, unable to persuade the Iranian regime to honor the Treaty of 1975 and settle the differences between the two countries by peaceful methods, decided to terminate both the Algiers Agreement and the Treaty of 1975 as a step to defend by force Iraq's territorial integrity. In a speech broadcast to the nation, he declared,

> The Iranian rulers' attitude, since assuming office, has confirmed their violation of the relations of good neighbourliness and their non-commitment to the clauses of the March Accord [Algiers Agreement]. They, therefore, fully bear the legal and de facto responsibility of rendering this Accord null and void. . . .
>
> Since the rulers of Iran have violated this accord . . . I here announce before you that the Accord of March 6, 1975 [Algiers Agreement] is terminated on our part too.
>
> Therefore, the legal relationship in Shatt al-Arab must return as it had been prior to March 6, 1975. . . .
>
> We affirm . . . that we seek good neighbourly relations with neighbouring countries, including Iran; that Iraq has no designs on Iranian territories and that we did not have any intention to wage war against Iran or expand the area of conflict beyond defending our rights and sovereignty.[11]

A note to this effect was sent to Iran on the same day. In the note, Iraq maintained that the Algiers Agreement and the Treaty of 1975 were terminated in accordance with the terms of both instruments (Paragraph 4 of the Algiers Agreement and Article 4 of the Treaty of 1975). Iraq, the note added, would call on the Islamic Republic of Iran to "accept the new situation and adopt a reasonable and wise attitude towards the exercise by Iraq of its sovereignty and legitimate rights . . . as the situation used to be prior to the aforementioned Algiers Agreement."[12] In other words, Iran was asked to surrender half the waters of the Shatt al-Arab, which it had acquired under the Algiers Agreement and the Treaty of 1975, and to recognize Iraq's sovereignty over the entire river, as it was defined under the Treaty of 1937.

The Iranian Foreign Office unexpectedly replied to this note in a letter sent not to Iraq, but to Kurt Waldheim, Secretary General of the United Nations (October 26, 1980). Iran rejected Iraq's "unilateral abrogation" of both the Algiers Agreement and the Treaty of 1975 and contested Iraq's argument that the violation of any single clause of either the Algiers Agreement or the Treaty of 1975 invalidated both instruments. The note also added that "Iran has always respected" the Treaty of 1975 and insisted that Iran "still considers itself bound by [its] provisions." It was Iraq, the note asserted, that had violated both the Algiers Agreement and the 1975 treaty by its aggression and subversive activities against Iran. Moreover, Articles 5 and 6 of the 1975 treaty (concerning the border between the two countries) were intended to be "inviolable, permanent, and final," according to the Iranian note. No provision in that treaty, Iran claimed, would permit unilateral abrogation. Thus Iran declared in no uncertain terms that the Algiers Agreement and the Treaty of 1975 were "still in force and binding."[13] To Iraq, this note indicated a reversal of Iran's position concerning the Algiers instruments, as expressed in earlier declarations.

In reply, Iraq sent two letters, one to Iran (November 16, 1980) and the other to the Secretary General Waldheim (November 26, 1980). In the former, Iraq reiterated its argument that Iran had in practice rendered invalid the Algiers Agreement and the Treaty of 1975 by the "persistent breaches of that Treaty not only through the declarations of Iranian officials that they do not recognize that Treaty but also through Iran's violations of its essential elements." In the letter to Waldheim, it was stated that Article 4, stipulating that the violation of one clause of the Treaty of 1975 renders the treaty as a whole null and void, and Article 6, providing procedures for resolving differences over borders, were not mutually exclusive, as "the application of Article 6 pre-supposes the existence of the Treaty through the non-violation of any of its indivisible elements." In other words, Article 6 can come into operation only on the strength of the validity of Article 4, but if Article 4 were violated, the whole treaty would be considered null and void as a package.[14] In several earlier notes, Iraq had invited Iran to negotiate their differences by negotiation, but its efforts were in vain. Thus Iran could hardly disclaim responsibility for the violation of the Algiers instruments when Iraq had called on Iran to settle their differences in accordance with the procedure laid down in those instruments.

Iran's reply to Iraq's termination of the Algiers instruments was to escalate frontier incidents and to interrupt navigation in the Shatt al-Arab. From September 10 to September 22, some fifteen incidents were reported, almost all in the Basra Province.[15] Several notes protesting Iran's actions were dispatched to Tihran, and in one of them Iraq warned that the "escalation of the use of violence" would put "full responsibility" on Iran, since Iraq could not possibly remain indifferent to violations of its territorial sovereignty.[16]

On September 21 and 22, 1980, the Iraqi army crossed the Iranian territory and began to shell urban centers that seemed to outside observers beyond the defence requirements of Iraqi security. Iraqi warplanes bombed airfields in Tihran and several other cities, and Iraqi forces advanced in several directions toward Dizful, Khurramshahr, and Susangird, northwest of al-Ahwaz. On the following day, heavy fighting continued, and the warplanes bombed several other urban areas. Iranian planes bombed the Iraqi oil installations at Faw and Port Bakr in the Basra Province, resulting in a complete stop of oil shipments. These military operations continued unabated, as the Iranian Revolutionary Army, resisting the advances of the Iraqi forces, seemed determined to counterattack. Thus the war began to drag on. Attempts by the United Nations and other third parties to persuade Iraq and Iran to call a cease-fire and turn to peaceful means of settlement were made from the very beginning of the war.

On September 21, Foreign Minister Saʿdun Hamadi of Iraq addressed a letter to Secretary General Waldheim for circulation among the United Nations members in which the steps taken by Iraq before resorting to force against Iran were explained. The letter, noted earlier, set forth the background of the Iraq–Iran conflict prior to the Iranian Revolution. For the reasons set forth in that letter Iraq decided to recover the three sectors by force and put an end to frontier violations.[17]

No sooner had hostilites between Iraq and Iran commenced than Waldheim brought the matter to the attention of President of the Security Council. After informal consultations among members, the president of the Security Council, considering the matter to "pose a grave threat to international peace and security," urged Waldheim to appeal to both parties for peace. But this was not all. Immediately after consultations, both the president of the Security Council and Waldheim addressed appeals to Iraq and Iran in which they were asked to "exercise the utmost restraint and to do what they could to negotiate a solution to their difficulties." The president of the Security Council, in another appeal on behalf of the members of the Security Council, asked the two states to "desist, as a first step towards a solution of the conflict, from all armed activity and all acts that may worsen the present dangerous situation and to settle their dispute by peaceful means."[18]

The first to reply was Iraq. In a letter to Waldheim, dated September 26, 1980, President Saddam Husayn stated,

Iraq has not interfered with the interests which affect the peace, security and economy of the world. . . . Iraq's objective is only to gain Iran's irrevocable recognition of Iraq's rights in its lands and sovereignty over its national waters. Is Iran prepared for that? . . . Iraq would wish to ask whether Iran is ready for a ceasefire. . . . This information is obviously important in considering our position.[19]

In acknowledging President Husayn's reply, Waldheim stated in a letter to Husayn that the request for a cease-fire had been made to Iran "in exactly the same terms" as it had been made to Iraq and that he was awaiting Iran's response to his request.

Before Iran responded to the appeals for a cease-fire, the Security Council invited Ismat Kittani, Iraq's representative to the United Nations, to participate, without a vote, in the discussion on the subject of the "situation between Iran and Iraq." Although the Iranian Permanent Mission to the United Nations informed Waldheim that the official response to his appeal for a cease-fire would arrive "at the latest by the morning of October 1, 1980," the Security Council held its first meeting on the Iraq–Iran conflict on September 28, before Iran's reply to the Security Council's appeal for peace had arrived.

On September 28, Ismat Kittani presented Iraq's case before the Security Council. He criticized Iran for failing to respect its treaty obligations and for attacking Iraqi territory. Since Iran refused to heed Iraq's warnings and settle its differences with Iraq by peaceful means, Kittani went on to explain, Iraq was bound to use force to put an end to Iran's violations of its treaty obligations.

The Security Council, following a discussion on the "situation between Iran and Iraq," adopted a resolution calling on both countries to refrain from the use of force and to accept the United Nations offer of a cease-fire and settlement of all differences by peaceful means. Waldheim was entrusted with the task of carrying out the Security Council Resolution and reporting within forty-eight hours.[20] It is of interest to note that the Security Council called the conflict between Iraq and Iran a "situation," refraining from calling it "war," in order to avoid a discussion of "aggression"—and, consequently, which nation was the aggressor—which might lead to the possibility of invoking punitive actions under Chapter 7 of the United Nations Charter. Nor have Iraq and Iran declared a state of war to exist between them. They have not even severed diplomatic relations; the embassy of each country is still officially in existence in the capital of the other, although the ambassador of each has been recalled, relegating his powers to a chargé d'affaires.

Waldheim, one day after communicating the Security Council Resolution to the parties concerned, received a reply from Saddam Husayn, in which he declared that Iraq was quite ready to "observe an immediate ceasefire," provided that Iran would abide by the cease-fire and enter into negotiations directly or indirectly "with a view to achieving a just and honorable solution ensuring our rights and sovereignty."[21]

Iran, however, rejected the resolution. In his letter to Waldheim, dated October 1, 1980, President Bani-Sadr did not state at the outset that the Security Council Resolution was rejected; he merely stated that he wanted to "clarify the position of the Islamic Republic of Iran concerning the present dispute with the Republic of Iraq." The events leading to the dispute between the two countries were set forth, as follows.

First, from the beginning of the revolution, Iraq had violated the Algiers Agreement by "sending Iraqi agents and armed units across our western and south-western borders . . . committing acts of sabotage and assisting counter-revolutionary groups." Moreover, Iraq "has been a haven" for Iranians from the former regime who became engaged in activities directed against the Islamic republic of Iran.

Second, Iraq expelled over 40,000 Iraq citizens either of Persian origin or of Shi'i persuasion in March and April 1980. This act, said Bani-Sadr, was an indication of Iraq's hostility toward Iran. "It is also a violation of human rights." It had been reported to the Secretary General, he added, but no condemnation of the act had been issued by any organ of the United Nations.

Third, before September 22, 1980, there were signs, such as the movement of Iraqi troops to Iran's borders, indicating that Iraq was already preparing to "escalate its hostile acts against Iran." Thus Iraq's act, said Bani-Sadr, was "premeditated" against Iran.

In view of these and other hostile acts, Bani-Sadr concluded, "the Security Council Resolution cannot be considered by our Government." In other words, Bani-Sadr made it clear that Iran was not prepared to accept the Security Council's call to settle its differences with Iraq by peaceful means, and its intent to continue the struggle obviously meant a rejection of the Security Council Resolution.[22]

Iran's rejection of the Security Council Resolution was even more emphatically stated by Ayat-Allah Khumayni, in a speech broadcast to the nation on October 3, 1980, in which he rejected any possibility of negotiations with Iraq and declared that the purpose of the war was not only to throw back the invaders, but also to punish "the criminal Ba'th leaders."[23] Ever since he made that speech, Khumayni had shown no sign that he would change his mind as long as the Ba'th leaders were in power.[24]

Despite Iran's rejection of the Security Council Resolution and Iraq's conditional acceptance (that Iran should abide by it and negotiate with Iraq directly or indirectly), the Security Council did not give up its efforts to persuade the two parties to accept a cease-fire. Since United Nations resolutions are recommendatory and cannot be enforced, except if the Security Council decided to enforce them by punitive measures under Chapter 7 of the charter, the Security Council's role was reduced merely to inviting Iran (October 5, 1980) to participate, without a vote, in the discussion on the conflict between Iran and Iraq, in an effort to get the two parties to accept a cease-fire.

The Security Council, presided over by Troyanovsky of the Soviet Union, met on October 17, 1980, to discuss "the situation between Iran and Iraq." Muhammad ʿAli Raja'i, Prime Minister of Iran, was invited to present Iran's case.

Raja'i, speaking in Persian, delivered a highly emotional speech. He said that his country was at "war" with Iraq—a war that had been imposed on it, he said later—and that he had just come "straight from the front," where "the spectacle of the dead and the wounded, which he had seen with his own eyes, would have moved the most heartless of men." In committing an act of aggression against Iran, Raja'i claimed, Iraq "has engaged 12 divisions and more than 2,500 tanks, as well as large quantities of weapons and hundreds of war planes" to kill innocent people, plunder and devastate the country, and even shell hospitals and schools, and kill schoolchildren and babies. He accused the Baʿth leaders—Saddam Husayn, in particular—of forcing this war on Iran in order to overthrow the Islamic republic and mutilate the revolutionary movement in Iran. Iraq's armies, which committed "inhuman acts" without mercy, were met with silence from the peoples of the world, who, claiming to "profess belief in dignity," declared their strict neutrality. He appealed to the "conscience" of the entire world, in particular to Muslim peoples, "with whom we share a common ideology and common values," to pass judgment on the acts committed against his country.

In his attempt to sway sympathizers, Raja'i complained that the United States and other Western or pro-Western states had "directly or indirectly been helping the Baʿthist Government of Iraq." "The United States with its AWACS aircraft in Saudi Arabia," he added, "controls the movements of Iranian troops and passes all information on to Iraq." Other states—including Egypt, Jordan, and Morocco—had supplied arms and munitions and spare parts to Iraq, he complained. Iraq's attitude toward Iran, he held, "is only a reflection of the hostility of the superpowers," and Iraq's aggression is but a part of the international aggression against the Islamic Revolution.

After his tirade against Iraq, the United States, and the Soviet Union, Raja'i turned to more specific charges against Iraq. First, he pointed out that the Algiers Agreement, considered by Iraq to be the cause of the conflict, was only a pretext for launching the aggression against Iran. Second, he charged Iraq with interference in the internal affairs of Iran. Third, he accused Iraq of assistance to hostile elements who used all material means at their disposal to undermine the Iranian regime. Two radio stations, he said, were very active on Iraqi territory. By inflammatory broadcasts, they incited opponents against the Iranian government. But the Islamic Republic, Raja'i said in no uncertain terms, "has never strayed from the terms of the Agreement," although it might have called for a review of the agreements already concluded with Iraq. He blamed all that had gone wrong between Iran and Iraq on the Ba'th leaders—in particular, Saddam Husayn, who "forced Imam Khumayni to leave Iraq" and sought to undo the work of the Iranian Revolution. But Raja'i assured the United Nations members that Iran, inspired by the ideology of Islam under the leadership of Imam Khumayni, would continue to fight the war and eventually "determine its own future."

In his final words, Raja'i warned that the Ba'th government had asked "for a ceasefire in order to deceive international public opinion." In rejecting that resolution, Raja'i said, "We wish to declare that a fair end to this war can be found only if the aggressor is vanquished and punished. That is our final position."[25]

Following Raja'i's trenchant attack on Iraq, Sa'dun Hamadi, the Iraqi Foreign Minister, on whom the president of the Security Council called, presented his country's views on "the situation between Iran and Iraq." Refraining from using abusive words, Hamadi simply said that Raja'i's "improper language" neither was in keeping with "the dignity of this meeting" nor would serve its "purpose." In defending the position of his country's president, whom Raja'i had described as a despot working against his own people, Hamadi said that "the President of Iraq rose from amongst the people, from the ranks of the very poor" and spent most of his life working for his people against "monarchy and dictatorship."

The rest of the speech was devoted to a discussion of the differences that had arisen between Iraq and Iran. He discussed, in particular, the questions of the termination of the Algiers Agreement and the Treaty of 1975, as well as other related matters, since he felt there was too much misunderstanding about them, partly because of differing interpretations of some clauses and partly because of a failure to fulfill some other clauses. Hamadi pointed out that Raja'i had accused Iraq of having violated both the Algiers Agreement and the Treaty of 1975 by terminating them unilaterally. In fact, he went on to explain, Iraq declared their termination after Iran had already terminated them by "word and deed," as he had pointed out in his speeches at the General Assembly and the Security Council of the United Nations. Iraq's action, he added, had been done in accordance with the Algiers Agreement

(Paragraph 4) and the Treaty of 1975 (Article 4). The Algiers Agreement, he explained, represented a package deal, the purpose of which was "to arrive at a final and permanent solution of the existing problems between the two countries," based on the principles of territorial integrity, the inviolability of frontiers, and noninterference in internal affairs. "The elements of the package deal," he stated, were (1) definitive demarcation of land frontiers, (2) delimitation of the frontier in the Shatt al-Arab according to the thalweg, and (3) reciprocal control of infiltration of a subversive character, no matter where it originated. Moreover, the two parties had agreed to consider all these obligations as indivisible, and "any violation of any one of them" would be considered contrary to the agreement.

With regard to the charge that Iraq had disregarded Article 6 of the 1975 treaty, which dealt with the differences arising from the interpretation and application of the treaty, Saʿdun Hamadi (who had represented Iraq during the negotiation and signing of both instruments) reiterated that the Treaty of 1975 was based on the Algiers Agreement and spelled out its technical details. In brief Saʿdun Hamadi pointed out that Article 6 became operative whenever either one of the two parties wished to invoke it, provided the treaty was still in force. But if Article 4 were violated, the treaty as a whole would no longer be in existence. "The continuous violations of the elements of the 1975 Treaty mentioned in Article 4," said Saʿdun Hamadi, "left Iraq with no Treaty to implement." When Iraq time and again reminded Iran of its obligations under the 1975 treaty, Hamadi added, Iranian leaders replied that the agreement did not meet Iran's interests and that Iran did not consider itself bound by it.

The Iranian Premier, said the Iraqi foreign minister, had criticized Iraq for interfering in Iran's internal affairs, but in reality Iran was responsible for the problems in which Iraq became involved. A case in point was the alleged Iraqi involvement with "the national minorities in Iran." Since the Iranian regime, said Hamadi, took steps to reformulate the state on a religious basis, stressing the tenets of the sect to which the majority of Iranians belong, its action naturally aroused the feelings of other sects in Iran, to which many Iraqis belong. If these minorities rise up and ask for national recognition, he said, that was not the fault of Iraq. "We in Iraq," he added, "have granted local autonomy and national and cultural rights" to national and religious minorities which Iran has failed to recognize.

Another case, Hamadi said, was Khumayni's departure from Iraq. When Khumayni was in Iraq, according to Hamadi, he tried to carry out subversive activities against the Shah, and invited foreign press and television correspondents to broadcast his attack on the Shah, which Iraq considered an interference in the internal affairs of Iran. After leaving Iraq, Khumayni seems to have considered his departure a personal insult.

Premier Raja'i, said the Iraqi foreign minister, had depicted Iraq as an aggressor. "Iraq," Hamadi maintained, "did not start the war with Iran." When Khumayni's subversive activities in Iraq failed to achieve his aim, Hamadi added, military action was resorted to. Indeed, Iraq had called the attention of the Security Council to the shelling of border towns and villages long before Khanaqin and Mandali were attacked by Iran on September 4, 1980. Despite Iraq's warnings, the shelling continued, which prompted Iraq to recapture the three central sectors that belonged to it. Iraq's action was met by intensive Iranian military operations. The violations of the Algiers Agreement and the Treaty of 1975, Hamadi said, prompted Iraq to declare these instruments null and void. Iran replied by heavy artillery shelling in populated areas within Iraq. Between June and September 1980, said Hamadi, the number of violations and military actions across Iraq reached 187. The Iraqi foreign minister mentioned several notes delivered to Iran, such as those of September 8 and 11, in which Iraq reiterated its desire not to widen the conflict and assured Iran that Iraq had no territorial ambitions, but Iran replied to none of these notes. Under these circumstances, Iraq declared the Algiers Agreement and the Treaty of 1975 terminated, since they had already been violated by Iran in word and deed. "Having exhausted all possible peaceful means," Hamadi said, "Iraq was left with only once choice, namely, the exercise of the right of self-defence for restoring Iraq's sovereignty over the totality of its territory." "We have," he concluded, "no territorial ambition in Iran, but we insist on the territorial integrity of Iraq, in land and water, and non-interference in our internal affairs."[26]

Further exchanges of views on specific matters were made by other members of the Iraqi and Iranian delegations (Riyad al-Qaysi for Iraq and Ardakhani for Iran), especially on such questions as interference in domestic affairs and political and ideological differences.

Following a discussion of the conflicting views of Iran and Iraq, the Security Council endorsed Secretary General Waldheim's endeavors to persuade the two parties to accept a cease-fire as a step toward a final settlement. Several subsequent resolutions, calling on both sides to refrain from the use of force and settle their differences by peaceful means, were passed by both the Security Council and the General Assembly. While Iraq has displayed a willingness to accept them in principle, Iran has remained uncompromising in rejecting them, has insisted that it was the victim of aggression, and has demanded condemnation of the Iraq regime. In other words, Iran will continue to resort to force until the Ba'th regime is overthrown. Meanwhile, missions by third parties, including some under the auspices of the United Nations, have been sent to Iran and Iraq to persuade them to settle their differences by peaceful means.

First, a fact-finding and good-will mission was established by the foreign ministers of Muslim countries at a meeting in New York (September 26, 1980), headed by President Zia al-Haqq of Pakistan, then president of the Islamic Conference Organization. The ultimate purpose of this mission was to establish a "peaceful settlement in the spirit of Islamic solidarity." The mission arrived in Tihran on September 27, 1980, and went to Baghdad on the following day. Meanwhile, Yasir ʿArafat, Chairman of the PLO, visited Tihran and Baghdad in support of al-Haqq's mission. This mission, coinciding with the Security Council's call for a peaceful settlement (September 28, 1980), came to nought. Since Iran had rejected the Security Council Resolution, al-Haqq saw no sign of readiness in the Iranian capital to accept a cease-fire. After returning to the United Nations, Zia al-Haqq, in a speech to the General Assembly (October 1, 1980), stated that "I have faithfully conveyed to each brother the views and position of the other in regard to a cessation of hostilities."[27] He also made public Iraq's offer of a conditional four-day cease-fire. Following the announcement of the unilateral cease-fire on October 4, 1980, Iraq declared that Iran had failed to observe the cease-fire and hostilities were resumed on the following day.

Second, the Muslim countries, not discouraged by the failure of the first mission to bring about peace, made another attempt a year later. In its meeting at Makka and Ta'if (January 25–29, 1981), the third summit of the Islamic Conference Organization, attended by all Muslim countries except Iran and Libya, appointed an enlarged permanent body called the Good-Offices Committee (Lajnat al-Masaʿi al-Hamida), composed of nine members and headed by President Ahmad Sekou-Touré of Gambia.[28] It was entrusted with the task of visiting the capitals of Iran and Iraq and preparing a set of proposals to be submitted to the two countries for their consideration. The presidents of Pakistan, Bangladesh, Gambia, and Guinea were included in the delegation, reflecting the seriousness of the mission. It arrived in Tihran on February 28, 1981, and was met at the airport by a large delegation that included President Bani-Sadr and Premier Raja'i. Following speeches of welcome and review of an honor guard, the mediating committee was informed that it would meet with the Supreme Defence Council, later in the day, and with the Ayat-Allah Khumayni, on the following day.

In the afternoon, a meeting of the Supreme Defence Council was held, and President Bani-Sadr said in his opening statement that the purpose of the meeting was to give the committee an opportunity to hear the council's views on the war. Sekou-Touré, in his address, stated that the purpose of the mission was not to act as a court, but to seek the "truth and justice." "Judgment on the existing differences," he pointed out, "would be a very difficult job." "The fact is," he went on to explain, "that the war has grieved us deeply and that the winner and the loser will in the end be losers in the view of Islam. Therefore this war should be ended through just and speedy measures. The devil separates the human beings from each other, but Islam unites them. . . ."

Sekou-Touré's remark about the devil prompted Bani-Sadr to answer with a few comments. He agreed with Sekou-Touré that "the devil exists." "Our request," he said, "is to decide who the devil is in this case." He asked the mediating committee to investigate "who the aggressor is." "Then after it is determined who the aggressor is," he added, "to punish him according to Islamic principles." In reply, Sekou-Touré said,

> We are not here in order to answer your questions, but rather to give you this message that we want peace and an end to this war. We implore you to go beyond touching on the problem and feel assured that no act of sacrifice and effort done in the cause of Islam can be called deceit. . . . This war is on no account limited to Iran and Iraq but rather other nations and other Muslim brothers are also suffering from the affliction touching them in this war. A peace between Iran and Iraq will undoubtedly make the Muslim Umma [nation] happy.

In this exchange of views, the last word seems to have been reserved to Bani-Sadr, who said, "To bring this war to an end, we propose: (a) Islamic decrees be enacted and the aggressor be identified and punished; and (b) as long as the aggresor remains in our territory we cannot agree to a cease-fire." [29] Although ʿAli Akbar Rafsanjani, speaker of the Majlis, did not participate in the discussion, he made known his answer to the mediating committee through statements published in the press: "As we are sure of victory and do not want to reward the aggressor, we refuse to negotiate." He added, "I think the best position for us is . . . to fight patiently until Saddam Husayn had dug his own grave in Iran and the Iraq nation is liberated." Indeed, the Majlis had already referred the question of mediation to a meeting of the Foreign Affairs and Defence committees, and their joint statement may be summarized as follows: (1) Iraq launched a military aggression against Iran, and "the reply to this aggression is war"; (2) as long as Iranian territory is under occupation "any talks . . . with the enemy is condemned and would be pointless"; and (3) the Iraqi regime enjoys no political legitimacy and its leaders should be "condemned and punished for their treason towards the two Muslim nations of Iraq and Iran." [30]

On March 1, 1981, the committee met with the Ayat-Allah Khumayni, hoping to discuss with him the ways to achieve a peaceful settlement. But

Khumayni, perhaps to avoid entering into an argument, delivered a long speech in Persian in which he asked the committee to decide on the basis of Quranic injunctions which nation was the aggressor in the war between Iran and Iraq. "If we are the aggressors," he said, "then fight us; if they [the Iraqi rulers] are the aggressors then fight them." In Islamic lands, he maintained, rulers should govern in accordance with the wishes of the people. If the people do not want them, he said, the rulers should resign. "If you ask the people of Iran and Iraq," he added, "they will tell you whether they want their governments or not." Khumayni asked the committee to hold a plebiscite in Iraq—provided it would be free—to determine whether the Iraqi people want to be governed by Saddam Husayn and his Ba'thist regime. "If they do not accept him," he pleaded, "how could Iran agree to make peace with a government whose people does not want it?" In his tirade against the Iraqi regime, Khumayni accused Saddam Husayn of oppression, injustice, and atheism. He asked the committee to investigate conditions and decide whether the rulers of Iraq were promoting the interests of their people or serving those of the United States and the Soviet Union. But, he concluded, if the rulers were to follow the teachings of Islam, not their simulacra, there would be no need for the 1 billion Muslims to agitate and make revolutions, as Islam provides guidance—through the belief in God—for prosperity and success. Before leaving Iran, the committee had no opportunity to discuss the question of a peaceful settlement with the Ayat-Allah, although it had not given up hope that a peaceful settlement might be ultimately acceptable to the Iranian leaders.[31]

After arriving in Baghdad on March 2, the committee met with President Saddam Husayn and his leading ministers. The committee was told that Iraq was ready to accept a cease-fire and was asked what Iran's attitude was. It replied that it had talked with President Bani-Sadr and others in high authority but had not yet received the official translation of Khumayni's address, although it knew the gist of it from the Persian media. The Iraqi leaders pointed out that nobody in the Islamic Republic of Iran can speak for Khumayni, and the committee had already realized that the question of cease-fire and peaceful settlement must be discussed with Khumayni himself.

After returning to Tihran on March 3, 1981, the committee met with President Bani-Sadr and attended a meeting of the Supreme Defence Council, at which Sekou-Touré said that "it was an honor . . . to carry out the second phase of their mission . . . and peaceful solution to end the war so that . . . the two countries of Iran and Iraq would be able to carry out their development plans." He stressed that his mission was "peace" and hoped that both sides would "put aside hatred, revenge and anger."[32] He then presented the council with the "package proposals" that the committee had prepared to end the war and asked that they be reviewed by the Iranian authorities. These proposals may be summarized as follows:

I. Principles
   1. Respect for the sovereignty of Iran and Iraq and their territorial integrity;
   2. Reaffirmation of the inadmissibility of the acquisition of territory by force;
   3. Reaffirmation of non-intervention and non-interference in internal affairs;
   4. Reaffirmation of the settlement of international disputes by peaceful means;
   5. Freedom of navigation in Shatt al-Arab;
II. Elements of Peaceful Settlement
   1. A ceasefire between Iraq and Iran shall come into force on the night of Thursday/Friday, March 13, 1981 at 00:00 hours;
   2. Withdrawal of Iraqi forces from Iranian territory shall commence from Friday, March 20, 1981, and will be completed within a period of four weeks, subject to consideration by a military sub-committee;
   3. Ceasefire and withdrawal shall take place under the supervision of Military Observers drawn from members of the Islamic Conference Organization acceptable to the two sides;
   4. The case of Shatt al-Arab be referred to a committee composed of members of the Islamic Conference Organization acceptable to Iran and Iraq to establish a final statute for this water-way;
   5. Negotiations for peaceful settlement of other disputes will follow withdrawal of Iraqi forces from Iranian territory;
   6. Exchange of declarations by Iran and Iraq of non-intervention and non-interference in the internal affairs of each other;
   7. Members of the Islamic Conference Organization will guarantee the observance by both sides of the commitments undertaken on the basis of the package peaceful settlement and, if necessary, maintain observers on both sides of the international frontiers for a certain period.
III. Interim Measures for Free Navigation in Shatt al-Arab
   1. From the time ceasefire comes into force until final agreement on Shatt al-Arab is reached, navigation in the water-way shall be organized and supervised by a special body under the auspices of the Islamic Conference Organization;
   2. This body may call on the Islamic Conference Organization to pro-

vide a peace-keeping contingent to insure free navigation in the Shatt
al-Arab during the interim period.

VI. The Islamic Peace Committee will establish a subcommittee to assist
the parties in implementing the provisions of the package proposals for
peaceful settlement.[33]

Shortly before the noon prayer, the committee met briefly with the Ayat-
Allah Khumayni and later performed the prayer, led by Khumayni himself.
After the prayer, Khumayni delivered a brief speech, in which he gave no
indication that he was prepared to change his hard-line stand before the
committee left Tihran in the evening for Baghdad. President Bani-Sadr, at
the airport to see the committee off, talked briefly about the war with Sekou-
Touré and Habib Shatti (Chatty). Although the committee felt that the Iranian authorities had taken a hard-line position, it hoped that the spirit of
Islamic brotherhood might eventually prevail.[34] On March 6, the Iranian
Supreme Defence Council rejected the proposals on the grounds that the
whole plan fell outside the 1975 agreements. Iran also demanded that an
international Islamic committee be appointed to investigate the question of
aggression.[35] By March 13, when the day set for the cease-fire to begin
under the "package proposals" was not observed, the committee came to
the conclusion that peace would not be achieved in the immediate future.

The committee, however, did not give up the hope of realizing peace by
an appeal to Islam and reason.[36] On October 22, 1982, the United Nations
General Assembly passed a resolution endorsing the endeavors of the Islamic Conference Organization to cooperate in the search for peace. This
resolution spurred the committee to renew its appeal to Iraq and Iran for
acceptance of the cease-fire, especially after Iraq had withdrawn its forces
from Iran. Iraq accepted the committee's offer, but Iran rejected it. Following the death of Sekou-Touré in 1983, Sir Dawda Kairaba Jawara, president
of Gambia and a member of the committee, became the chairman of a newly
reorganized Good-Offices Committee. He proposed to visit Iran and Iraq and
seek again to arrange a cease-fire. Iraq welcomed his visit, but Iran agreed
to receive him only as president of Gambia, not as chairman of the Good-
Offices Committee. In the circumstances, Jawara could not proceed with his
plan, and the activities of his committee came to a standstill. Nevertheless,
the Islamic Conference Organization continued its efforts by official and
unofficial contacts to press for a cease-fire and settlement by peaceful means.
Both Iran and Iraq were invited to the Islamic summit meeting in Kuwayt
(January 26–28, 1987), but Iran refused to attend on the grounds that Kuwayt was not a neutral country. The Islamic Conference Organization could

do nothing beyond renewing the offer of a cease-fire and settlement of the dispute by peaceful means.

Finally, two other mediating missions whose members were drawn from countries outside Islamic lands attempted to achieve a settlement by peaceful means. As early as November 1980, Secretary General Waldheim, in accordance with the Security Council Resolution of September 28, 1980, announced that a special representative had been appointed to visit Iran and Iraq and explore the possibilities of a peaceful settlement. He invited Olof Palme, former prime minister of Sweden, to undertake the mission, and Palme agreed to serve in that capacity. Following approval of the mission by Iran and Iraq, Waldheim informed the president of the Security Council (November 11, 1980) and Palme left for Tihran a week later. Palme visited Iran and Iraq five times (once in 1980, three times in 1981, and once in 1982), and met with high-ranking officials in both countries.

Palme saw willingness to discuss fundamental issues of the war but found that there was neither flexibility on the combatants' part nor any change in their views from those already presented by their representatives to the United Nations. In Iraq, Palme met with the highest ranking men in the regime—President Saddam Husayn, Deputy Prime Minister Tariq 'Aziz, and Foreign Minister Sa'dun Hamadi—and all presented their views with clarity and agreement. But in Iran, he found differences of opinion between civilian leaders, represented by President Bani-Sadr, and clerical leaders, represented by Ayat-Allah Beheshti and Hujjat al-Islam Rafsanjani. In his last visit (1982), he met with President 'Ali Khami'ini and Prime Minister Raja'i. Because of rivalry and a struggle for power between the two groups, Palme had to conduct his discussions separately with each group, and it was difficult to obtain consensus on all their views.

Palme's method of dealing with the conflict between the two countries was to divide the issues into two categories: first, the fundamental principles of International Law governing the issues; second, specific proposals concerning all issues that would be topics for negotiations.

With regard to the general principles, Palme found general meeting of the minds of both sides on

1. Noninterference in domestic affairs.
2. Respect for sovereignty and territorial integrity.
3. Refrain from resort to force.
4. Freedom of navigation in the Gulf.

During his second and third visits, Palme discussed the following proposals:

1. Cessation of hostilities.
2. Withdrawal of forces to lines coinciding on the whole with the borders drawn under the 1975 treaty.
3. Negotiations covering all issues arising from the 1975 treaty, such as the thalweg in the Shatt-al-Arab.

The two sides seem to have taken such inflexible positions on some issues that it was impossible to reconcile them. The Iranians demanded the withdrawal of all forces before they would enter into any negotiations, and the Iraqis insisted that withdrawal be included in the negotiations after the acceptance of a cease-fire.

During his negotiations with both sides, Palme tried to assist in resolving the question of foreign vessels that were trapped in the Shatt al-Arab, unable to sail out. He proposed a limited cease-fire to allow the clearing of the Shatt al-Arab of sunken boats and unexploded ammunition. Apart from the question of sovereignty over the river, a tentative formula was reached to enable the vessels to be removed. Disagreement on the financing of the operation frustrated final agreement. Iran proposed to pay half the cost, while Iraq proposed to pay the full cost. Alternative sources were explored, which led eventually to the removal of the vessels, but Palme left before the matter was resolved.

In his final visit to Tihran (February 27–28, 1982), palme came to realize that the political will for a negotiated settlement did not exist. Although he did not give up the hope of achieving peace, as the Secretary General had informed Iran and Iraq, the mission was prorogued *sine die* after the death of its head.[37]

The other mediating committee was appointed by the Non-Aligned Movement (NAM) in 1981 because both Iran and Iraq were members of that movement. The committee, established at the ministerial level, was composed of the foreign ministers of Algeria (later withdrawn after Iraq's objection), Cuba, India, Pakistan, Gambia, and Malaysia, and a representative of the PLO. After visiting Tihran and Baghdad, the committee found no willingness to negotiate, save that each combatant demanded the expulsion of the other from the NAM. Iran opposed Iraq's hosting the Non-Aligned Conference in Baghdad in September 1982. President Husayn, probably because of war conditions and security considerations, agreed that the conference should be held elsewhere. Further visits in late 1981 and early 1982 were also unsuccessful. Finally, when it was bluntly told by Iran in 1982 that the conflict with Iraq could be decided only on the battlefront, the mission dissolved.

It might be useful at this stage to discuss the legal position of each side and the conditions set forth to bring the war to an end. True, Iraq has accepted peaceful means of settlement and Iran has rejected them, but each has laid down its own conditions for settlement.

What were these conditions? At the outset, when the war had just started, Iraq's demands were expressed in broad terms, as stated by President Saddam Husayn in his speech to the nation on September 28, 1980.

> What we demand is that the Iranian Government should explicitly recognize, both de facto and de jure, the historical and legitimate rights of Iraq over its land and waters; adhere to the policy of good neighbourliness; abandon its racist, aggressive and expansionist tendencies; give up its vicious attempts to interfere in the internal affairs of countries in the region; return every single inch of land it had susurped from the Homeland, view its rights and those of the Arabs and Iraqis on this basis, and respect international laws and conventions.[38]

Apart from the broad terms, the specific Iraqi demands may be summarized as follows. First, the land and water frontiers referred to by President Husayn were those defined and determined in various historical treaties and arrangements, which were finally embodied in the Treaty of 1937 and the Algiers instruments of 1975. Since the latter treaties recognized the thalweg as the water frontier, it was repudiated by Iraq on September 17, 1980, on the grounds that the Algiers Agreement and the 1975 treaty had been violated by Iran. Iraq thus demanded the water frontier to be the eastern bank of the Shatt al-Arab, as it had been before March 6, 1975 (that is, prior to the Treaty of 1975).[39]

Second, Iraq demanded that the islands of Abu Musa and the two Tunbs, to which Saddam Husayn referred in his speech as part of the Arab Homeland, be returned to the United Arab Aimrates, as Iran had annexed them without the consent of their owners in 1971.[40] Some high-ranking Iraqi officials hoped that the Arabs of al-Ahwas Province (referred to by Iraq as Arabistan) might be emancipated from Iranian rule. Had they staged an uprising and put forth such a demand, Iraq might have recognized the de facto existence of this province as an Arab entity. It was perhaps owing to such expectations that the Iraqi forces penetrated beyond legitimate frontier claims, which prompted President Carter to declare that the Iraqi forces had gone "beyond the ultimate goal" (presumably the limits set under the Treaty of 1975) and that the United States "would like to see [the Iraqi] invading forces withdrawn."[41] Iraq no longer raises Arabistan and the Gulf islands as issues for negotiation.

Third, Iraq demanded respect for the principle of noninterference in internal affairs, a principle that Iran had accepted under the Treaty of 1975 and the United Nations Charter. Khumayni's doctrine of "exporting the Revolution,'' endorsed by the Islamic Republic of Iran, is obviously inconsistent with the principle of noninterference in other nations' domestic affairs, as it had already stirred Shiʿi followers in Iraq and other Gulf countries to agitate against their own governments.

Fourth, Iraq demanded respect for the principle of good-neighborly relations (embodied in several treaties between the two countries) and the settlement of disputes by peaceful means. Iran has not only refused to negotiate with the Baʿth regime, but also demanded the change of this regime before it would negotiate with Iraq.

Fifth, the legal and diplomatic rationale for Iraq's crossing of Iran's frontier was presented by Foreign Minister Saʿdun Hamadi before the United Nations General Assembly on October 3, 1980, and the Security Council on October 15, 1980. In order to recover the three central land sectors, Hamadi asserted, ''my Government was left with only one choice, namely, the exercise of self-defence for the purpose of restoring Iraq's sovereignty over the totality of its territory.'' ''There was,'' he added (borrowing words from a well-known but controversial case [*The Caroline*, 1837]), ''a necessity of self-defence, instant, overwhelming, leaving no choice of means and no movement of deliberations.'' In other words, Iraq had resorted to the so-called preemptive act, often invoked in recent years, to justify what otherwise is considered an aggressive act.

Sixth, Iraq offered to submit its dispute with Iran to arbitration or any other kind of judicial procedure, in accordance with treaties existing between the two countries and the principles and rules of International Law. Although Iran reproached Iraq for the termination of the Algiers instruments, it has rejected Iraq's request to the advisory opinion of the International Court of Justice on the matter.

Iran, rejecting resort to all means of peaceful settlement, claims that Iraq's initial attack on its territory, on September 21, 1980, was an act of aggression for which Iraq's rulers were solely responsible, although Iran's claim has been contradicted by Iraq's argument that Iran had already resorted to violence when it attacked villages and towns within Iraqi territory on September 4, 1980. True, Iraq's attack was massive and disproportionate were it to be considered a retaliation for Iran's earlier resort to violence. But Iran's incursions across the frontier were continual and showed no sign that they would stop, despite repeated protests by Iraq. At any rate, Iran's argument lost much of its strength when Iraq accepted the Security Council Resolution calling for a cease-fire and withdrew its forces to international frontiers in June 1982. Iran not only turned down the Security Council's calls for withdrawal to international frontiers, but also invaded Iraqi territory and has occupied some areas since 1984. Moreover, Iran has demanded the identification and punishment of the aggressor, implying that Iraq was the aggressor, and a change of regime before it will negotiate with Iraq. What would be the form of the regime agreeable to the Iranian leaders? Since Iran has continued to advocate the doctrine of "exporting the Revolution," the only regime acceptable to it would be a twin of the Islamic Republic in Iran. Iraq's confessional and political structure, accordingly, would have to be completely altered to conform to the Islamic Republic in Iran. No sovereign state would accept such a demand, as it not only is contrary to the United Nations Charter and the principles governing the relationships among nations, but also is humiliating and insulting to the people of Iraq. Iran has also demanded the payment of war reparations (reiterated on several occasions, especially by ʿAli Akbar Rafsanjani, speaker of the Majlis, on October 22, 1982).

Richard A. Falk, arguing that since both Iraq and Iran have violated International Law, suggested that "an essential ingredient of any successful settlement of this [Iraq–Iran] War is an arrangement in which neither side appears to have lost."[42] Although certain aspects of the war, such as border conflicts and the treatment of persons with dual nationality, are legal disputes, others, such as the relationship between state and religion and the Sunni–Shiʿa controversy, are political disputes. The legal disputes might be dealt with after a cease-fire is established. In a political settlement agreeable to both sides, however, there is merit in arguing that neither side should appear to have lost, a subject to which we shall return in the final chapter.

# IX

# The Iraq–Iran War: The Ideological Aspect

The undeclared war between Iraq and Iran has been fought on several fronts. We have already seen how it was fought in international councils by diplomacy before the forces of the two countries met on the battleground. There is still another front, which may be called the "war of words," stressing essentially its ideological aspect. This kind of warfare is fought not by armed forces, but by broadcasts over the media, confessional controversies, and, not infrequently, subversive forays. Unlike guns, which destroy physical targets, the war of words strives to influence the minds and attitudes of citizens by invoking stereotypes and distorting news in order to enhance one side's position and undermine the other's. The war of words may not have the immediate effect of a decisive battle, but it might have a permeating influence on the ultimate outcome of the war.

At the outset, the war of words was not waged on specific issues—frontier dispute and intervention in domestic affairs—but on the nature of the regime and the character of the leaders who exercised authority in each country. In the press and propaganda literature, the superiority of the Islamic form of government was stressed and texts of Khumayni's writings were widely circulated and translated into foreign languages. The clerical leaders claimed that the Islamic Republic set up by the Iranian Revolution is the ideal form of Islamic government, which the Prophet Muhammad and his successors had laid down. This government, as expounded in Khumayni's teachings, is opposed to oppression and aims at the achievement of justice in accordance with Islamic law. The Iranian leadership wished to establish this kind of government not only for believers in Iran, but also, according to Khumayni's doctrine of "exporting the revolution," for believers in other Islamic lands.

The two regimes—the Ba'th of Iraq and the Islamic republic of Iran—were by their very nature prepared to engage in ideological warfare even before the hostilities commenced. Both aspired to establish new social orders, one embodying the fundamental principles and values of Islam and the other espousing modern doctrines such as nationalism, socialism, and democracy. The two ideologies, however, though seemingly poles apart, have certain goals in common—the assertion of independence, opposition to foreign pressures, and noninterference in domestic affairs. Since the great majority of the population in both countries is Muslim, Islam became the source of inspiration for both; the Iranian leadership regarded it as the very foundation of its regime, and the Baʿth party as a component of its national ideology.

We have already set forth Khumayni's ideas about government based on the assumption that ultimate authority belongs to God and that its exercise was delegated to the Imam. In the absence of the Imam, divine authority was and still is exercised in theory by the mujtahids, although in practice it was divided; civil authority was exercised by the Shah and religious authority by the collectivity of the mujtahids.[1] Khumayni rejected the viewpoint that the Imam's authority was divided into civil and religious spheres and held that it should be exercised by a council of jurists (wilayat-i faqih) in an acting capacity. His views, considered the basic source of the new Iranian constitution, have been widely endorsed, and those opposed to it were discredited and outlawed by the revolutionary regime.

The Shiʿi form of Islamic government, based on the doctrine of the legitimacy of the Imam, obviously asserts a more intimate relationship between the religious and the political authorities than does the Sunni creed, based on the principles of public consent and consultation. Since Iraq has long been governed by Sunni leaders, the traditions of public consent and other secular elements are more deeply rooted in the country than in the Shiʿi community of Iran. Small wonder that the existing political system in Iraq reveals Sunni patterns of authority, although Shiʿi followers have often challenged Sunni leadership and asserted Shiʿi principles of authority.

The leaders of the Islamic Republic in Iran, although often speaking in the name of Islam in general, have consciously sought to promote their own particular brand of Islam into a favored position. Iran's stress of Shiʿism provided ample ammunition for the press and other media to engage in ideological warfare by inciting Shiʿi followers in Iraq to denounce the Baʿth regime as biased against Shiʿi of non-Arab descent. The extremists have gone so far as to deny the legitimacy of the Baʿth government on religious and historical grounds. The Iranian zealots thus betrayed not only narrow confessional and parochial propensities, but also a misunderstanding of tradition and history.

The Baʿth party, according to Iranian propaganda, came into power not by a legitimate Islamic upheaval, but by violent and illegal military actions. The Iranian leaders maintained that just as Mu'awiya, founder of the Umayyad dynasty, had usurped power by force from the Calip 'Ali, the first Imam, and passed it on to his son Yazid in A.D. 680, so Hasan al-Bakr, President of Iraq, had come into power by a military coup (1968) and passed on the seals of office to Saddam Husayn (1979). Although the Iranian leaders must have known that Saddam Husayn did not succeed Bakr by designation, since he was enthroned by the hierarchy of his party, they probably wished the Shi'i world to be reminded of the historic event of the usurpation of authority by Mu'awiya from the Imam 'Ali, depriving his descendants of the right to succession. The Shi'a in Iraq were thus implicitly told that authority, usurped by Sunni leaders in a country whose majority are Shi'is, belongs to them. Since it was taken for granted that in Iraq power could not be gained by peaceful methods, Shi'i followers in Iraq should seize it by revolution, just as their co-religionists had done in Iran.[2]

We have already noted how the Shi'i community played a meaningful role in the politics of Iraq before the revolution of 1958 and how some of their leaders reached the highest positions in the regime. But, in reality, only residual powers were taken over from Sunni hands. In 1958, following the overthrow of the old regime, the new regime failed to set up a viable government acceptable to all major groups, irrespective of confessional division. For over a decade, the political system, dominated by the military, paid little or no attention to confessional sensitivities. Shi'i followers thus became restless, and the clerical leaders began to claim religious authority. Since its rise to power in 1968, the Ba'th party has sought to reduce confessional tensions by proposing the establishment of a new political system based on the principles of nationalism, socialism, and parliamentary democracy, rather than on religious affiliation.

By asserting national principles, the Ba'th leadership was able to attract an increasing number of young men by promising them equal opportunities to play a significant role in the party's hierarchy and in high government offices. As a result, Shi'i leaders, whose influence was rising since the July Revolution of 1958, were opposed to the Ba'th regime. When the mujtahids in Iran succeeded in gaining power, the clerics of Persian descent in Iraq became very active and tried to persuade those of Arab descent to join them in demanding the establishment of an Islamic government. The appetite of the Iraqi Shi'i leaders for power was thus whetted by their peers in Iran who, in accordance with Khumayni's doctrine of "exporting the Revolution," offered to assist them in their endeavors to achieve power.

Long before the Iranian mujtahids were able to overthrow the Shah's regime, Shi'i clerics in Iraq had been concerned about their leadership slipping from their hands, particularly during the Qasim regime when a swelling tide of young Shiʿi followers were drawn to the Communist party. Sayyid Muhsim al-Hakim, the Shiʿi spiritual leader, who had become quite concerned about the spread of Communism in the country, began to criticize the authorities for granting Communists freedom of political expression.[3] The Shah of Iran, disenchanted with the military regime in Iraq, appealed to al-Hakim and his followers to join him in a campaign against Communism, although the Shah was not on good terms with the mujtahids in Iran, who were opposed to his oppressive regime. Despite the Shah's hypocrisy in dealing with the religious leaders in Iran and Iraq, al-Hakim saw in the cooperation with the Shah an opportunity to oppose Communism as a means for the rehabilitation of religious leadership in both countries.

The arrival of the Ayat-Allah Khumayni at Najaf in 1965 had a far-reaching effect on the shifting attitude of the Iraqi Shiʿi leaders, first toward the Shah and later toward the Baʿth regime. At the outset, Khumayni consciously assumed a low profile, at least in part in deference to al-Hakim, who enjoyed much prestige in the Shiʿi community, although he and al-Hakim held entirely different views about the role of religious leadership in politics.[4] Perhaps even more important is the fact that Khumayni, as an exiled guest, was not expected to participate in a propaganda campaign against the Shah, since the Iraqi regime, presided over by ʿAbd al-Salam ʿArif, maintained a fairly good working relationship with Iran. Nor did Khumayni then enjoy the position of high religious authority, although he was acknowledged as a prominent mujtahid, since the marjiʿiya (high religious authority) belonged to such illustrious men as Muhammad Riza Kulbaykani, Shariʿat-Madari, and Hasan Qummi in Iran, and to al-Hakim, al-Khawʿi, and others in Iraq. For almost three years, until the ʿArif regime was overthrown by the Baʿth party, Khumayni's activities were confined largely to teaching and personal contacts with a relatively limited number of friends and disciples.

Khumayni's position in Iraq began to improve following the coming of the Baʿth party to power in 1968. Since the Shah was not on good terms with the Baʿth government, he became involved with opponents of the Baʿth (to whom he supplied weapons) in an effort to overthrow the new regime in 1969. To their great satisfaction, Khumayni and his followers were thus encouraged to speak openly against the Shah and were offered the unrestricted use of the media to agitate against him from their sanctuaries in the country. He and his son Mustafa, who died under mysterious circumstances (in which the Shah, it is alleged, was involved), were in continual contact with high-level politicians and were provided with all the means for agitating against the Shah. It was at this time, especially after al-Hakim's death, that Khumayni's star as a religious leader began to rise, and his followers grew in number. Indeed, Iraq became a haven for fugitives from Iran.[5]

Even before Khumayni's fortuitous flirtation with the Ba'th regime had started, there were several organizations devoted to the improvement of the Shi'i position in the country. As noted earlier, the activities of those organizations were spurred in reaction to the July Revolution of 1958, which granted freedom of expression to various ideological groups. At the outset, the new generation, irrespective of religious identity, sympathized with the aims of the July Revolution, hoping that the new regime, promising freedom and prosperity, would live up to its promises. When, to their surprise, the young men found that the principal leaders had betrayed the revolution and become engaged in a struggle for power, most young Shi'i followers began gradually to fall back on their Shi'i identity and to seek the guidance of religious leaders. Their disenchantment became even more pronounced with 'Arif, well known for his Sunni bias, restricted their activities. Under the Ba'th regime, Pan-Arab slogans, although acceptable to young Shi'i men, alienated almost all older Shi'i followers and drove them into the fold of traditional Shi'i leadership.

The ideological warfare between the Islamic Republic and the Ba'th regime began when the Iranian mujtahids were persecuted in Iran under the Shah and a number of them found protection in Shi'i holy centers, where they rallied supporters against the Shah's regime. Since the revolutionary trends in Iraq ran contrary to Shi'i aspirations, the chasm between the Shi'i and Sunni communities began to widen. Following the Iranian Revolution, Shi'i followers in Iraq looked to Iran for support. The goal of Iran's ideological warfare was and still is to enable the Shi'i community in Iraq to play a meaningful role in the country's political system and, ultimately, to assume its leadership. To achieve this aim, the revival of Shi'i teachings and the preparation of a new generation that would undertake leadership were deemed necessary. Before the Iranian Revolution, the stress was on nonpolitical organizations, but after the Islamic republic was established, more active organizations began to work, sometimes resorting to violence.

In 1965, Hizb al-Da'wa al-Islamiya (the Islamic Call party) came into existence. At the outset, its goals were not political, as its activities were confined to the revival of Shi'i teachings and restricted to areas where Shi'i followers outnumbered Sunnis, such as Najaf, Karbala, Samarra, and Kazimayn. But its ultimate goal was unmistakably political, since the stress on Shi'i teachings tended to raise tensions, especially among the rank and file. Confessional tension, especially in Shi'i strongholds, often led to street incidents and violence. The police were bound to intervene to keep the peace, but the police action spurred a controversy among religious leaders of both communities, and thus the friction and high emotions that began at the lower level often turned into a more sophisticated debate at higher levels, revealing the subterranean confessional feeling in the country. The Da'wa's teachings were thus indirectly responsible for the revival of sectarian strife.

The Da'wa sought to attract Shi'i followers to the fold, as the Shi'i young generation was still deeply involved in the Communist movement. Its attitude toward al-Hakim, who was on good terms with the Shah and on fairly good terms with the Ba'th regime, was ambivalent. Some members, sharing al-Hakim's concern about Communism, paid little or not attention to the Shah's relations with Iraq; others, especially the mujtahids who had fled Iran, warned al-Hakim's followers about the Shah's hypocrisy in his flirtation with the Shi'i leadership in Iraq. Still others, like the Ayat-Allah al-Khaw'i, although opposed to the Shah, took a neutral position in deference to al-Hakim. Only after al-Hakim's death in 1970 did the Da'wa begin to turn against the Shah, in response to the complaints of an increasing number of mujtahids who had left Iran.

In 1970, the high religious authority in Iraq passed from al-Hakim to Muhammad Baqir al-Sadr. This event may well be regarded as a turning point in the attitude of the Da'wa toward both the Shah and the Ba'th regime. In 1975, when the Ba'th regime came to terms with the Shah, Khumayni finally parted company with the Ba'th and began to support the Da'wa party. He also supported the new spiritual leadership of al-Sadr, whose ideas he shared, and the two leaders won each other's confidence when Khumayni was still in Iraq. Following his return to Iran and the establishment of the Islamic Republic, Khumayni is reported to have invited al-Sadr to lead a revolution in Iraq in order to establish an Islamic government.[6]

But before discussing al-Sadr's relations with the Iranian Revolution and its impact on Iraq, a few words about Sadr's background and ideas might be illuminating. He was born into an Arab family in Najaf in 1930. The dedication of his family to Shi'i affairs ever since its founder came to the country over a century ago was well-known in the Shi'i world.[7] Several members of the family, drawn into public affairs, distinguished themselves in nationalist activities. One of them, Muhammad al-Sadr, took an active part in the Iraq Revolt of 1920 and won a high reputation in nationalist circles. After the establishment of the Iraqi national regime in 1921, he was drawn into politics, becoming first a senator, in 1926, and later the prime minister, in 1948.[8]

Muhammad Baqir al-Sadr, growing up in a family milieu of scholarship and piety, showed an interest in learning while still young. He published several books and articles on Islam covering a wide range of subjects. His works, intended for the general reader, may not be very original and highly scholarly, as they summarized essentially the traditional Islamic teachings. Outwardly, the object of his works was to reform society, irrespective of confessional differences; but the ultimate goal was obviously political and critical of the regime, by making clear that the alternative programs of re- form offered by modern secular doctrines—Pan-Arabism, socialism and oth- ers—which the Baʿth have advocated, would eventually lead to the revival of confessional tensions and other forms of discrimination and injustice. Nor was Sadr merely a writer on abstract subjects; he was also an activist who distinguished himself as a spiritual leader concerned with the welfare of the community and the progress of Islamic communities in other countries. In recognition of his personal qualifications—charisma, scholarship, and fluency in speech—and the prestige of his family in the service of the state and the community, his authority as chief marjiʿ to act on behalf of the Imam during his absence (ghayba) was acknowledged by other ulama, and his claim was confirmed by Shiʿi followers. As chief marjiʿ, he had the power to issue legal opinions on questions of the day that would be binding on the com- munity as a whole.[9] The ʿulama in various ranks were expected to carry the messages of the high religious authority to the community of believers on all levels. The rank and file—most of whom were illiterate and in poor conditions—would dutifully respond to the call of the high authority and act accordingly.

In accordance with the Shiʿi creed in Iran and Iraq, the twelfth Imam had gone into occultation, but will eventually return to restore power and justice to his followers. The Shiʿi community had been warned that before his re- turn it might suffer persecution and injustice under oppressive rulers (tughat). During the Imam's absence, the mujtahids provide guidance on all matters of concern to the community. Under such a system, political decisions are not expected to be determined solely by social and economic considerations, but by the mujtahids in accordance with the dictates of religion and law (Shariʿa). If social and economic considerations were in conflict with the decisions made by the Imam—in his absence the existing high spiritual au- thority (al-marjiʿiya)—these decisions would be binding on all Shiʿi follow- ers. However, the spiritual authority of the mujtahids is often exercised by more than one acknowledged high authority, each claiming to guide Shiʿi followers in his town or area. All other mujtahids (whether in his town or another), often called the muqallids (who follow the high authority), are expected to carry out those decisions through their contacts with the com- munity.

One of the effective instruments with which the mujtahids carry out the call of the high religious authority is the religious educational institutions (al-hawzat al-ʿilmiya). In recent years, there has been a noticeable increase in the number of these institutions as well as in the number of teachers who went to Iraq from Iran, presumably on the grounds that a growing number of Shiʿi followers had fallen under the influence of Communist and other secular teachings that undermine not only the high religious authority, but also the position of the Shiʿi community as a whole. From Ottoman days, the Shiʿi community began to establish these educational institutions because the Ottoman Porte discriminated against the Shiʿa and refused to admit Shiʿi young men into their schools or employ them in government service. Shiʿi followers thus were bound to fall back on their own communal organizations and schools, run by the ulama and ultimately guided by the high religious authority. Students from Iran often went to Najaf, as its schools had an excellent reputation in scholarshisp throughout the Shiʿi world, and Iranian scholars who studied at Najaf often taught in Shiʿi schools in both Iraq and Iran.[10] Brought up in confessional schools, Shiʿi teachers and students from both countries not unnaturally fraternized and developed an affinity for one another and a common identity before the era of nationalism, based on the principle of fealty (Walaya) to the Imam, which in practice meant loyalty to the chief religious authority, regardless of whether he was an Arab or a Persian, or whether he resided in Najaf (Iraq) or Qum (Iran). Small wonder that when the war broke out and the media spread the call of the mujtahids in Iran to their co-religionists in Iraq to rise up against the atheist Baʿthist regime, the initial response was almost spontaneous, leading to high tension and attempted uprisings in Shiʿi strongholds.[11]

Closely connected with confessionalism in the ideological warfare is nationalism. Both the Baʿth regime in Iraq and the Shah's regime in Iran stressed national symbols of identity, as nationalism was considered the foundation of their rule. But the revolutionary government in Iran, asserting Islamic standards, declared that it was opposed to the Shah's regime for the very fact that it had invoked national symbols, to which Islam is opposed because nationalism discriminates among believers—considered brothers in religion—irrespective of descent and country. Islam is opposed to all forms of national identity because national values are exclusive and must be superseded by universal values.[12]

The press and other media, echoing Khumayni's ideas, denounced time and again the Baʿth regime as secular and racist because it discriminated among believers on the grounds of national and ethnocultural identities. Underground organizations, inspired by Iran, became very active in the country. They spread propaganda extolling the merits of Islamic government and called on believers to overthrow the Baʿth regime and establish an Islamic regime that would recognize no distinction among believers.[13] Both the Daʿwa party and the Islamic Amal conducted a campaign against the Baʿth leadership, attacking in particular Saddam Husayn and the "anti-religious" regime over which he presided. In one of the circulars issued by the Da'wa, it was announced that the ulama had come to an agreement that the Ba'th regime must be overthrown, and the Iraqi people were called on to replace it with an Islamic government.[14]

The Iranian regime has waged ideological war through the press and subversive activities not only in Iraq, but also in other Arab countries and the West. At the outset, Iraq did not have an adequate public-relations service to cope with Iranian propaganda, which depicted Iraq as an aggressor aiming to overthrow the Islamic republic and dismember its territory. Although Iraq sought to settle its differences with Iran by peaceful methods, the Iranian press continued to brand Iraq as an aggressor, even after its forces had withdrawn from Iran in 1982. It was not until the Iranian forces launched frontal attacks inside Iraqi territory and bombed urban centers (especially in the Basra Province and the Kurdish settlements in the northern provinces) that the Western press began to pay more attention to Iraqi broadcasts and complaints, although the Iranian propaganda literature continued to extol the merits of the Iranian Revolution and to undermine the Iraqi regime.[15]

To justify its attacks on Iraqi territory, the Iranian leadership has claimed that they are dictated by the divine orders of the hidden Imam, for whom the Ayat-Allah Khumayni has become the spokesman. The Iranian campaigns have been labeled by symbolic names; the most widely known are the Karbala campaigns No. 1, 2, and so on (Karbala is the site where al-Husayn, Imam 'Ali's second son, fell in battle while struggling to take over authority from Sunni usurpers), obviously intended to inspire believers to follow in the footsteps of Husayn to recover Shi'i holy lands that are still in Sunni (Ba'th) hands. Such campaigns have not only inspired young men to take the field—if they fall in battle they will be rewarded in Paradise—but have also attracted Shi'i followers in other Gulf countries to participate in subversive activities.

The Iranian leaders have been able to inspire believers outside Iran as well as to enter into alliances with leaders of nationalist movements to support their cause. Moreover, their public statements against Israeli occupation of Arab lands and their support of the PLO have provided President Asad and Colonel Qadhdhafi with pretexts for their support of Iran, since Iraq's preoccupation with the war with Iran has undermined the general Arab position vis-à-vis Israel by keeping the forces of both countries locked in the war instead of joining the Arab armies in a confrontation with Israel. Likewise, Iran has supported Kurds in their struggle with Iraq to achieve nationalist goals, while its propaganda literature, stressing Islamic standards, criticizes Iraq for asserting national identities. In Lebanon, the Iranian leaders have supported subversive activities, irrespective of national identity. In their pursuit of power, the terroristic methods of Iran's protégés have degenerated to nothing less than a license for violence and abuse of Islamic values.

But this is not all. In their ideological warfare with Iraq, the Iranian leaders sought to play on the emotions of the people and create a wedge between rulers and ruled. They declared on more than one occasion that Iran and the Iranian people have no ill-feeling toward the Iraqi people and that they wish to maintain brotherly and good-neighborly relations between them. But such relation, the Iraqis were told, would be possible to maintain only after the Baʿth regime was replaced by another—a government based on the Islamic principles of law and justice and ready to promote peace and order in the region. The assertion of Islamic principles irrespective of confessional differences is obviously intended to negate the accusation that Iranian Shiʿism, embellished with Persian traditions and practices foreign to Islam, amounts to a different creed. Needless to say, the Iraqi leaders have also assured the Iranian people that they have no enmity toward them. They offered them peace and security and called on their leaders to settle their differences with Iraq by peaceful means.[16]

The Iranian leaders have sought the support not only of sympathizers outside Iran, but also of Iraqi emigrés in Iran. Some had already left Iraq before the war began, but a greater number had been expelled by the Iraqi government. Other Iraqis were in Syria and Libya, where they were given asylum by the authorities. At the outset, the exiles in Iran and Syria sought cooperation with members of the Daʿwa party and the Amal organization, but the leaders of these organizations held that since their objective was to establish an Islamic government in Iraq, leadership of that government should be entrusted to clerical and not to civilian hands. As a result, two organizations emerged, one in Iran and the other in Syria. The latter, with headquarters in Damascus, was supported by the Syrian authorities. Called the Revolutionary Islamic and National Front of Iraq, it had both lay and religious members and a civilian leadership, headed by Hasan al-Naqib, a retired Iraqi general since 1981. The other organization was formally established in 1982, and its clerical leadership was entrusted to Hujjat al-Islam Muhammad Baqir al-Hakim, who became its president in 1986.

Hakim, an Iraqi of Persian descent, held dual nationality. He is thus at home in both Arab and Iranian societies. Possessing no special qualifications for leadership save his membership in the well-known Hakim family, he has become an instrument in the hands of Iranian authorities. Nor has he commanded the respect of his more than eighty followers, representing diverse parties, groups, and houses. Hakim's organization is not, strictly speaking, a government-in-exile, as its members wish to call it, but a council, assisted by an executive committee and a secretariat with several units attached to it. The principal function of the organization is to look after the welfare of refugees and other exiles from Iraq. But its ultimate aim is to replace the Ba'th regime with an Islamic republic that would establish close cooperation between the two countries. To achieve this goal, it has become completely dependent on the Iranian government in seeking supporters by clandestine activities and the publication of the *al-Shahid* and *Liwa al-Sadr* newspapers, *al-Jihad* magazines, and other propaganda literature. It had been hoped that once the city of Basra had fallen to the Iranian forces, Hakim and his followers would move their headquarters into that city to carry out a more intensive propaganda campaign against the Iraqi regime. Its activities, however, have been discredited in Iraq, since it played the role of a fifth column for the enemy.

How did the Ba'th regime respond to Iran's ideological warfare? At the outset, it took several drastic steps to overcome internal subversive activities, to which we have already referred. Perhaps the most important was the withdrawal of the Iraqi forces from Iranian territory in 1982, giving evidence that Iraq had no territorial ambition beyond the defence and security of the country. This step, carried out in accordance with United Nations resolutions since 1980, confirmed Iraq's commitments to settle its differences with Iran by peaceful means. As a consequence, Iran bears responsibility for the continuation of the war, since it has refused to accept all calls for a cease-fire on the grounds that Iraq started the war. But in reality, Iran wishes to go beyond that. It has already demanded replacement of the Iraqi regime with another, presumably presided over by leaders agreeable to it, before it would negotiate with Iraq. This demand, apart from raising serious legal and diplomatic questions, is an insult to the people of Iraq, who are being asked to accept a regime dictated by foreign pressure. Moreover, were the Iraqi regime to be replaced by one similar to the Islamic regime in Iran, confessional tensions would be aggravated.

Before the Iraqi forces were withdrawn from Iran, the Shi'i community in Iraq, especially the Shi'a of Arab descent, were in a difficult position. As Iraqi nationals, they were bound to support their government, but as members of the Shi'i community, they were expected to sympathize with their co-religionists whose country had been subjected to Iraqi attacks. However, since the Iraqi forces have been withdrawn, Shi'i followers of Arab descent have often called on the Iranian leaders to settle their difference with Iraq by peaceful means.

The Shi'i followers of Persian descent who held dual nationality fall into an entirely different category. Even before hostilities began, most of them seem to have been engaged in a propaganda campaign against the Ba'th regime. They often called on their Arab co-religionists to join in the campaign against the regime. Aware of their activities, the Iraqi government at first warned them against participating in subversive activities. Later, after hostilities began, it deported over 40,000, according to Iranian sources.[17] Although some who held dual nationality had been born in Iraq and were assimilated by long residence and intermarriage, their deportation was carried out as a measure of internal security. The spiritual leader of these dissidents, Muhammad Baqir al-Sadr, who held dual nationality, was brought to trial and summarily executed on April 24, 1980, on the grounds that he had issued a fatwa justifying subversive activities against the state.[18] By the end of 1981, the revolutionary movement had been suppressed, and almost all those engaged in subversive activities had been hunted down and thrown into prison. Most of those who fled the country joined al-Hakim's organization, which had already been engaged in subversive activities against Iraq.

In its pursuit to overcome dissent, the Iraqi regime followed a pragmatic procedure. It made its position crystal clear on the question of religion and politics. It assured all who would refrain from political opposition that they had nothing to fear, were they to attend to merely their religious affairs, as the exercise of religious duties is a private affair. But it warned against meddling in politics under the guise of religion. This policy, called in Arab parlance the targhib and tarhib (reward and punishment, or "carrot and stick") proved a constructive approach to a very sensitive question.[19]

But this was not all. The Ba'th government announced that its program of reconstruction and development would be continued despite the exigencies of war. In particular, the principal Shi'i centers—including Najaf and Karbala—received the special attention of the authorities, and the ulama who were prepared to cooperate with the regime were rewarded. The regime's readiness to extend all possible facilities and to stress national above confessional loyalties impressed many Shi'i followers, who realized that their interests would be best served by cooperating with, not opposing, the regime. After the withdrawal of the Iraqi forces from Iran, an increasing number of Shi'i leaders began to realize that it was in their own interest to support the regime, since they saw no valid reason for Iran's continuation of the war.

The Iranian Revolution has given rise to a sensitive issue about the relationship between the Iranian and Iraqi spiritual leaderships. Najaf, home to the most reputable centers of learning in the Shi'i world, has suffered a decline in prestige and in student attendance. Some of the most distinguished mujtahids from Iran—Burujirdi, Kalbaykani, Kashani, and Khumayni—had at one time or another studied and taught in its institutions. In recent years, Qum and other centers, where Persian mujtahids have become very active, began to compete with Shiʿi centers of learning in Iraq. In Najaf, the situation has become even more sensitive, not only because its position has been virtually eclipsed by Qum, but also because the high spiritual leadership has slipped from the hands of its ulama. Since in the Shiʿi community, there is no official hierarchy—each mujtahid attains his spiritual authority and prestige by his own endeavors—some of the Shiʿi spiritual leaders in Iraq were quite prepared to voice their disagreement with the spiritual leaders in Iran, just as some distinguished spiritual leaders in Iran— such as Shariʿat-Madari and others—have disagreed with Khumayni's own interpretation of religious doctrines. The Baʿth leaders, aware of this delicate situation, have encouraged some of the Shiʿi leaders of Arab descent—including Kashif al-Ghita, ʿAli al-Saghir, and the Bahr ul-ʿUlum brothers (Husayn, Jaʿfar, ʿIzz al-Din, ʿAla al-Din, and Hasan)—to challenge Khumayni's spiritual leadership.[20] No less important is the readiness of the younger ʿulama, tempted by ambition to follow in the footsteps of their predecessors in Arab nationalist activities, to speak out against their peers in Iran in an effort to enhance the influence and prestige of the Iraqi Shiʿi centers. These local and parochial sensitivities were rationalized on creedal differences. For example, the Iranian mujtahids have stressed certain rituals and practices introduced into Shiʿi teachings by the Safawi dynasty, such as self-flagellation during the Ashura mournings and the cursing of the early Caliphs who succeeded the Prophet, whose validity has been questioned by the Iraqi ʿulama. The Baʿth regime, in support of the Iraqi Shiʿi leadership, has encouraged the ʿulama to discuss their differences with the Iranian mujtahids and to persuade Shiʿi followers to give up their support to the Iranian regime.[21]

The Gulf War is thus not merely a battle between the armed forces of two sides, but also a confrontation of manifold dimensions between two ways of life. One, claiming to derive its inspiration from religion, stresses doctrines not at all compatible with the modern age. yet if it is ever to survive, it is bound to make concessions and accommodate to modern conditions of life. The other, aspiring to achieve rapid progress and development, has yet to adjust material and technological innovations to the cultural heritage that the people not only in Iraq, but also in all other Arab countries, still honor.

# X

# Involvement of the Gulf States in the Iraq–Iran War

The Iraq–Iran War, beginning as a conflict between two Gulf neighbors, is no longer confined to only two countries. Today, it is a war in which almost all the Arab countries are directly or indirectly involved, and it has profoundly affected their relations with other regional and global powers to varying degrees. It is true that the conflict between Iraq and Iran (and earlier between Persia and the Ottoman Empire) was the product of historical forces with which the other Gulf countries had little or nothing to do, but its impact as a war fought on land, on sea, and in the air could not possibly be isolated or ignored by other Gulf neighbors. Perhaps no less important is the fact that almost the whole Gulf has become a theater of war, although only the specific area of some fifty miles around the Kharj Island, declared by Iraq a "war zone," and most of Iran's southern Gulf territorial waters, declared by Iran "exclusive," have been prohibited to navigation by noncombatant vessels. It has, therefore, aptly been called the "Gulf War."

To what extent, it may be asked, has each Gulf country been involved in the war?

Before we discuss the involvement of each Gulf state, perhaps the aims of the two major countries and their attitudes toward the other Gulf countries might be summarized. Iran's message to Iraq and its neighbors—indeed, to all other Islamic lands—has already been made clear: it aims to establish an Islamic government in each that would enforce Islamic law and deal with domestic and foreign affairs in accordance with Islamic standards. Whether each country would be formally associated with Iran in an equal or a subordinate relationship is not made clear, but Iran would probably aspire to a privileged, if not the predominant position in the house of Islam as *primus inter pares*. Above all, were Iran, the major Gulf state, to win the war, it would be in a position to play the role of leader in the maintenance of peace

117

and security, especially if the Gulf were exposed to the threat or intervention of a foreign power.

Iraq's relations with other Arab Gulf states have not always been smooth and friendly, owing partly to dynastic rivalries but mainly to Iraq's territorial ambition, particularly in Kuwayt, which has alarmed other Gulf countries. Following the fall of the monarchy in 1958, the establishment of the republican regime in Iraq and the recurring military coups did not in themselves pose a threat, but the subsequent adoption of socialism and other radical doctrines by the Baʿth party, let alone its alliance with the Soviet Union, have aroused concern in high Gulf circles. The alarm deepened in recent years as the Baʿth regime, stressing Pan-Arab goals, has shown greater interest in Gulf affairs than ever before.

In an attempt to mollify their Arab Gulf neighbors, the Baʿth leaders have time and again assured Gulf rulers that they have no intention of interfering in their domestic affairs, least of all of undermining the ruling dynasties. They tried in vain to impress on them the need to follow an Arab Gulf policy aimed in the main at fostering harmony and cooperation to oppose foreign intervention and to limit immigration of non-Arabs to the Arabian coast of the Gulf, which had the effect of de-Arabizing the character of the Gulf. While the Arab states shared Iraq's concern in principle, they were suspicious of Iraq's ambition to play the role of leader in Gulf security affairs. For this reason, it was suggested that an understanding between Iraq and Saudi Arabia was deemed necessary before embarking on an Arab Gulf policy. The Iraqi leaders seem to have been confident that their policy would eventually meet with approval, as it would promote general Arab interests and not only Iraq's own special interests. They must have also realized that such a policy could evolve only slowly and cautiously by means of direct negotiations, and it should not be imposed as a ''grand design'' embodying ideological goals.

Negotiations between Saudi Arabia and Iraq began when the Amir Fahd, then Crown Prince, visited Baghdad in 1974. A year later, Saddam Husayn, Vice Chairman of the Revolutionary Command Council, visited Riyad. In both visits, the Iraqi leaders were told in all candor that a number of pending issues were awaiting resolution before the larger political question of Arab Gulf policy would be tackled. Saddam Husayn, who seems to have gotten along rather well with the Amir Fahd, made it crystal clear that Iraq's policy aimed at strengthening Arab independence and territorial integrity and not at threatening them, despite its alliance with the Soviet Union. Under no circumstance, Saddam Husayn added, would Iraq allow foreign pressures or radical doctrines inconsistent with Arab nationalism to penetrate into Iraq's territory. These initial talks were followed by the settlement of a number of pending issues—the dispute over the neutral zone and delimitations of the northern frontier between the two countries (July 2, 1975) and other matters—which greatly helped to smooth relations and to clear the atmosphere of suspicion.

Egypt's unilateral decision under Anwar al-Sadat's leadership to deal with Israel, considered as another attempt to impose a peace settlement of the Arab-Israeli conflict without Arab consent, brought Saudi Arabia and Iraq closer than ever. The ninth Arab summit meeting, held in Baghdad (in November 1978—presided over by President Ahmad Hasan al-Bakr, the Amir Fahd head of the Saudi delegation, in cooperation with Saddam Husayn, the head of the Iraqi delegation—succeeded in reaffirming Arab solidarity. Two new events brought about an even closer cooperation between the two countries: the Soviet invasion of Afghanistan and the Islamic Revolution in Iran. The first was perceived as an initial step by a superpower to threaten the Gulf region, the other, as a more immediate threat to their regimes. Three months after the Baghdad Summit, on February 8, 1980, Iraq proclaimed the Arab National Declaration, consisting of eight points, in which it committed itself to refrain from resorting to force, save in legitimate self-defence, in its relations with Arab and other regional countries. The eight points may be summarized as follows:

1. The rejection of the presence or the facilitation of the presence of foreign armies and military bases or any facilities in any form in the Arab homeland.
2. The prohibition of the use of force by one Arab country against another. All disputes among Arab countries shall be settled by peaceful methods.
3. Clause 2 shall apply to the relationship between an Arab country and any other neighboring country; no force shall be resorted to save in self-defence.
4. Solidarity among Arab countries shall be maintained in the event of an attack against the sovereignty and territorial integrity of any Arab country.
5. The affirmation of Arab commitments to observe all the laws and conventions governing the use of water, air, and land in the relationship with any country that is not at war with an Arab country.
6. The adoption of a neutral and nonaligned stand in any conflict or war in which Arab sovereignty and territorial integrity are not involved.
7. The adoption of constructive plans leading to joint Arab economic cooperation and Arab unity.
8. Iraq would enter into discussion with any other Arab country prepared to achieve the goals embodied in this declaration. The purpose of this dec-

laration is not to supersede but to reinforce the Arab League Pact and the Treaty of Joint Arab Defence and Economic Cooperation.[1]

Iraq's role at the Baghdad summit meeting and its announcement of the Arab National Declaration, aiming at achieving greater solidarity and cooperation among Arab countries, enhanced its position in inter-Arab affairs. But the "spirit of Baghdad," violated by Syria and Libya, did not last very long. Above all, the establishment of the Islamic revolutionary regime in Iran, focusing its frontal attacks against Iraq before any other country, diverted Iraq's attention from Arab to Iranian affairs.[2]

No sooner did hostilities between Iraq and Iran begin, in September 1980, than Iraq declared that it was fighting not only its own war, but an "Arab war," and that it was defending not only its territorial integrity, but the integrity of Arab lands as a whole from an impending Persian domination.[3] But neither did Iraq seek the support of other Arab countries nor were the other Arab countries prepared to join in the fighting.[4] At the outset, Iraq seems to have been determined to win the decisive battle—often called "Saddam's Qadisiya"—against Iran single-handed. But the expected quick victory of Iraq proved illusory, and the tug of war dragged on. When, however, Iran began to counterattack and seemed to be gaining ground, Iraq's Arab neighbors could not remain indifferent to the possibility of Iraq's losing the war.

After the outbreak of hostilities, the Arab Gulf countries were not adequately prepared to cope with their own security problems, and the oil installations on the western coast of the Gulf were exposed to attacks by Iranian aircrafts. Outwardly, most Arab countries seemed sympathetic to Iraq; but some hesitated to commit themselves, while others took a stand against it. Their initial postures were perhaps nowhere made more clear than at a meeting of the Arab League in Tunis. In their first meeting immediately after the outbreak of the war (September 22, 1980), the members of the Arab League arrived at no agreement or consensus. The Arab Gulf countries, supported by Jordan, Morocco, Mauritania, and North Yaman, took the side of Iraq. Syria and Libya, each for its own reason, supported Iran. Algeria, prepared again to offer its good offices to Iraq and Iran (as it had done in 1975), maintained a strict neutrality. Despite this division, Shadli Qulaybi (Chadli Klibi), the Secretary General, announced that the Arab League Council called for a cease fire and the normalization of relations between Iraq and Iran on the basis of the noninterference of each in the domestic affairs of the other. Iran, however, warned its Arab neighbors by official and unofficial channels—as well as by violent signals—to stay out of the war.

The first Arab country to declare with great enthusiasm its support of Iraq was not a Gulf country, but one far from the theater of war—Jordan. On learning the news of the outbreak of the Iraq-Iran War, King Husayn began at once to ponder its impact on the Arab world and what his country's response to it should be. It did not take him very long to realize that it was not merely a conflict between two rival regimes—one religious, and the other antireligious—as the Iranian media announced to believers in Islamic lands and to the outside world. It was, he perceived, a more serious event that might well have far-reaching consequences for the whole Gulf region. Indeed, it was a confrontation between a new hegemonic and revolutionary regime, raising the banner of Islam to exploit religious sentiment in Islamic lands, and an Arab nationalist regime, representing in one form or another the general Arab national pattern of polity that began to develop in northern Arab lands and the Nile Valley after the First World War.

King Husayn considered the Arab Revolt, which his great-grandfather King Husayn Bin ʿAli, the Sharif of Makka, had led during the First World War, the fountain spring of the Arab nationalist movement that subsequently spread throughout the Arab world. He maintained that the Arab Baʿthist regimes in Iraq and Syria, although quarreling at present over leadership and the form of unity they aspire to forge, are supporters of the same Arab nationalist movement prevailing today in Jordan and other Arab lands. In other words, he held that the Islamic Revolution is a challenge to the Arab nationalist movement, which aspires to establish a modern and progressive united Arab national state. What would happen to this movement, he speculated, were Iraq to lose the war? Most Arab leaders and their regimes, including all the Arab Gulf states, had already been denounced, it will be recalled, as antireligious and traitors to Islam. Were the Islamic regime in Iran to win the war, the Iranian Revolution would sweep not only the Arab regime in Iraq, but also all other Arab regimes in the Gulf and elsewhere. They would be replaced by governments similar to the Islamic republic in Iran, in accordance with Khumayni's doctrine of "exporting the Revolution." Moreover, King Husayn has reacted not only to the challenge to his country, but also to the attack on him as an infidel and a traitor to Islam. For King Husayn is a great-grandson of the Prophet Muhammad through the lineage of ʿAli's first son, al-Hasan (ʿAli being the first Imam to Shiʿi followers), although Khumayni also claims to be a great-grandson of the Prophet through the lineage of ʿAli's second son, al-Husayn. King Husayn takes pride in the fact that the Arab caliphs (successors to the prophet) provided the leadership for the establishment of the Islamic Empire and that his great-grandfather al-Husayn Bin ʿAli, the Sharif of Makka, provided the leadership for the Arab nationalist movement in the tradition of the early successors to the Prophet.

King Husyan's own ideas about Arab nationalism, he maintains, are also derived from Islam, and he argues that, contrary to Khumayni's views, there is no contradiction between Islam and nationalism.[5]

Against this background, it is understandable why King Husayn reacted adversely to the Iranian Revolution when he first learned the news. On September 24, 1980, two days after they learned of the outbreak of war, Husayn and Prime Minister Mudar Badran left Amman for Baghdad by car. Even before they left Amman, the Jordanian army seems to have been placed on alert, and Jordanian airfields were offered to Iraqi aircraft as safe shelters. In Baghdad, King Husayn and Prime Minister Badran met with President Saddam Husayn and offered their country's support not only politically, but also militarily and otherwise. Appreciative of the offer, Saddam Husayn pointed out that Iraq was not in need of military assistance, but he discussed other matters in which the two countries could cooperate.

It is true that King Husayn may have felt safe offering his services to Iraq without inhibition, as Jordan is relatively remote from the theater of the war. But he was quite prepared to dispatch a military force at the beginning, although Iraq made no such request. In 1982, when Iran began to attack Iraqi territory, several hundred volunteers, calling themselves the Yarmuk Force, went to Iraq.[6] Perhaps even more important was Jordan's assistance in obtaining arms for Iraq after the Soviet Union, Iraq's chief supplier, had declared its neutrality in the war. From China, Spain, and other Western European countries, Jordan was able to purchase, on behalf of Iraq, a considerable quantity of arms at a time when Iraq was in real need of tanks, ammunition, and spare parts for the Soviet arms it had obtained before the war. Jordan also assisted in the negotiations for obtaining loans and credit from Western countries for its purchases of war matériel.

Above all, Jordan extended economic assistance by throwing open its Red Sea outlet through the Port of ʿAqaba for Iraq's export and import trade as well as all other land and air facilities. Before Iraq entered into agreements with Turkey and Saudi Arabia to export its oil, Jordan was perhaps the only country open to Iraq as an outlet to the outside world; Kuwayt extended limited facilities because it was exposed to danger. Before the war, Iraq had entered into several agreements with Jordan providing for economic, commercial, and cultural cooperation, and some of these agreements were revised to meet the new exigencies of the war. A case in point is the agreement concluded in 1975, concerning the use of ʿAqaba as an outlet for Iraq's imports and exports. This agreement, confined at first to about 1.5 million tons of goods annually, was extended to over 5 million tons annually after the war started. Shortly before the war began, some 130 Iraqi commercial vessels carrying Iraqi goods were on their way to Iraq via the Gulf. Instructed to unload the cargo at ʿAqaba, they returned in time before they were exposed to danger in Gulf waters. ʿAqaba is still Iraq's only outlet to the high seas, and Jordan sought to maintain free navigation in the Red Sea.

But this was not all. Jordan rendered invaluable service in its mediation with other Arab countries that were reluctant to support Iraq either because of fear of retaliation by Iran or because of rivalry with the Iraqi leadership, such as Syria and Libya. But to these mediations we shall return later.

No less significant were King Husayn's personal endeavors through his many visits to Arab and Western capitals to explain Iraq's position in the war and defend it against attacks that it had started the war. At the outset, Iraq was, indeed, in need of a friend well known to European and Western leaders who could advise them of Iraq's intentions and readiness to settle its differences with Iran by peaceful means. These endeavors enhanced Iraq's position, particularly during the first two or three years of the war, when Iraq's own public relations had not yet been given adequate attention by European and Western media. In all these capacities, King Husayn has become the eloquent spokesman for Iraq in Arab and Western councils.[7]

Outwardly, the attitude of Iraq's Arab Gulf neighbors seemed rather ambivalent. No public statements about their official positions had been made, even though their sympathy may have been with Iraq, mainly because their eastern coastal cities and oil fields were exposed to the danger of bombing and destruction. They may also have not expected the war to last very long—indeed, there was an expectation that it would be an easy victory for Iraq—mainly on the grounds of chaotic conditions in Iran and disarray in the army, as noted earlier. As the war began to drag on and Iranian resistance stiffened, the Arab Gulf countries began to have second thoughts about the outcome of the war. Before the Gulf Cooperation Council (GCC) was established in 1982, consultations among Arab Gulf leaders about security measures and how to bring the war to an end were continually conducted, both individually and collectively. But no clear consensus about strategy and means to end the war emerged. With regard to security requirements, the Arab Gulf countries may be divided into three categories.

First, some advocated the use of diplomacy, whether unilaterally or collectively, as a means to contain the conflict and reduce its impact from spreading into the whole Gulf region. Each country, it was argued, should take the necessary measures in accordance with its resources and domestic requirements. If needed, it was held, two or more countries might cooperate in a limited way insofar as their geopolitical and strategic requirements were concerned. Kuwayt, Qatar, and the United Arab Amirates, to mention but three of the small countries, seem to have taken this position, although each pursued its own diplomatic procedure. This category may be called the self-help countries.

Second, other countries, viewing the war from the beginning as a more serious threat to their existence, held that diplomacy by itself would be inadequate to contain the war or reduce its impact on the region, let alone to bring it to an end. To make diplomacy more meaningful, they advocated greater dependence on military preparedness. ʿUman, advocating this method more forcefully than others, saw in the Iranian Revolution not only a threat to the Gulf region, but also, perhaps even more important, a pretext for Soviet penetration of or intervention in the region, particularly if the Islamic Republic were to come to an understanding with the Soviet Union or were replaced by a Communist regime.

Third, most countries, however, saw in the two foregoing approaches no contradiction and advocated a combination of the two as the surest method for safeguarding territorial integrity and regional security. Saudi Arabia, consisting of a large portion of Arabia extending from the Gulf to the Red Sea, has taken this position, as its defence and security requirements necessarily depended on diplomacy and military strength. Indeed, the experiences of most nations have demonstrated that military strength is always necessary to augment diplomatic procedures.

While the debate on security measures was still going on, there was discussion in high political circles (in both Saudi Arabia and other Arab Gulf countries) as to which of the two countries—Iraq or Iran—would present a greater threat to the region were it to win the war. Some, especially in Saudi Arabia, reflecting the views of traditional religious groups, held that were Iran to win the war, perhaps a conciliatory attitude toward the Islamic Revolution might enhance the regime's reputation and reduce religious tension.[8] For example, the Shiʿi disturbances in Bahrayn and in eastern Arabia, spurred by the Iranian Revolution, prompted high-level officials to consider how to respond to the Iranian challenge. While some advocated resorting to force, others urged that conciliatory measures be carried out. We shall discuss later which of these two responses proved more effective.

A number of leaders, in contrast, were concerned about the possible revival of Pan-Arabism, reminiscent of the Nasir era, were Iraq to win the war and assert its Baʿthist ideology. When the war began to drag on, however, and Iran insisted on continuing the struggle even after Iraq had withdrawn its forces from Iran, Arab Gulf leaders became quite alarmed about the intentions of the Iranian leaders, were Iraq to lose the war. Since all Arab Gulf leaders had been denounced as secular and antireligious—hence traitors to Islam—they began to wonder what their fate would be were the

clerics to seize power in their own countries, in view of the merciless on-
slaught against opponents in Iran. Against this impending peril, real or
imaginary, what security measures did the Arab Gulf countries take?

Saudi Arabia, the leading Arab Gulf country, did not make a public state-
ment about its official position, but King Khalid, in a telephone conversation
with Saddam Husayn on September 25, 1980, pledged support for Iraq. The
King's oral message may be taken to reflect the sympathy of the Saudi
Arabian outlook, and the Iranian leaders were not unaware of the situation.
The Saudi leadership, however, looked at the war neither from the strict
Arab outlook nor from the narrow viewpoint that it was a conflict between
two neighbors. True, Iraq is an Arab country with which Saudi Arabia is
bound by the national identity of Arabism, but the Saudi leaders also looked
at Iran as a neighboring Muslim country with which Saudi Arabia is bound
by the broader religious identity of Islam. For these and other considerations, regional as well as global, the Saudi leaders regarded the war from
two perspectives—Arab and Islamic—and sought to bring it to an end as a
matter of concern to all believers in Arab and Islamic lands.

From the Islamic perspective, Saudi Arabia is the cradle of Islam and the
home of the two holy sanctuaries of Makka and Madina, in the direction of
which believers all over the world pray five times a day and to which they
make an annual pilgrimage. Thus the Saudi leaders were bound to cultivate
friendly relations with all Muslim countries, including Iran, and sought to
raise the question of the Gulf War at almost every Islamic gathering as a
matter of concern to all. They also deemed it necessary that they be impartial in their relationships with all Muslim countries if they were to have a
significant influence in Islamic councils to bring the war to an end as soon
as possible.

As an Arab country, however, Saudi Arabia sought to play the role of a
big brother in the family. It had already repaired ruptured relations between
Syria and Jordan and had tried on more than one occasion to iron out differences among other states, particularly between Iraq and Syria. But the
Gulf War, a conflict between two Muslim countries, one Arab and the other
non-Arab, proved too complicated to bring under control. Under pressure
from Arab countries, it was bound to tilt toward Iraq, as the position of the
whole Arab family would be undermined were Iraq to be crushed. In taking
such an ambivalent position toward Islam and Arabism, the Saudi dynasty
has not been spared the reproach of critics. But Saudi Arabia has refused to
abandon its policy of moderation and balance, especially its friendship with
the West, despite occasional criticism. It has thus prudently used its diplomacy and resources to build up its military strength and has invited other
Arab Gulf countries to cooperate in the pursuit of peace and security.

With regard to their territorial integrity and security, most other Arab Gulf countries followed, in the main, the Saudi policy: reliance on diplomacy augmented by military buildup. Each country but one, relying on its own resources, sought before the establishment of the Gulf Cooperation Council to cooperate with Saudi Arabia through bilateral agreements; the exception was Kuwayt, which sought Saudi cooperation without entering into formal agreements. For Kuwayt, surrounded by the three major Gulf states—Iran, Iraq, and Saudi Arabia—is obviously the weakest militarily; it therefore felt bound to depend primarily on diplomacy, drawing on its past experiences in this field. ʿUman, holding that diplomacy is an inadequate means for defence, stressed dependence on military strength and sought the assistance of Western powers, partly because of the subversive activities of its next-door neighbor—South Yaman—and partly because of possible Soviet intervention via the Indian Ocean. The other Arab Gulf countries—Bahrayn, Qatar, and the United Arab Amirates—situated on the lower Gulf between Kuwayt and ʿUman, are somewhat remote from the area of military operations, and the degree of their dependence on diplomacy or military power varied from one to another. Before the establishment of the Gulf Cooperation Council, these countries often discussed security measures only in reaction to immediate violent acts.

Even before the Iraq-Iran war broke out, the countries closest to Iraq—Kuwayt, Saudi Arabia, and Bahrayn—had been exposed to Iranian threats. On June 12, 1980, three months before the war started, explosions occurred at the office of *al-Raʿy al-ʿAm (Public Opinion),* a daily Kuwayti paper known for its criticism of the Iranian Revolution. On June 13, three Iranian planes violated Kuwayti airspace, but Kuwaiti anti-aircraft guns seem to have forced them to retreat.

Bahrayn, where the Shiʿi community is relatively well organized and active, reacted more quickly to Iranian instigations. On April 24, 1980, almost a week after Muhammad Baqir al-Sadr, the Shiʿi spiritual leader, had been executed in Iraq, Shiʿi followers began to demonstrate in protest against the Iraqi action. One of the demonstrators, detained by the authorities, died while in custody on May 10 (it was alleged that he had died from torture). The incident prompted further demonstrations, leading to the arrest of other demonstrators, and the tension in the country was aggravated. Like Iraq, Bahrayn followed the policy of "carrot and stick," reflecting the differing views of high-level leaders, some demanding resort to force and others, reliance on diplomacy. The Shiʿi leaders, as well as other Sunni religious and liberal groups, had presented petitions to the authorities in the early 1970s. The religious groups demanded the literal application of Islamic law; others called for parliamentary democracy, individual freedom, release of political detainees, and termination of the American military presence. These rather ambitious demands were shared by other elements who maintained that political development should be in line with the social and economic progress of the country. As a result of popular pressure, elections were held in 1972 for the establishment of a national assembly and the promulgation of a constitution promising freedom and social reforms. From 1973 to 1975, Bahrayn had a short-lived parliamentary government, which was suddenly suspended by the authorities on the grounds of internal instability and national security. The ruler of Bahrayn, Shaykh Isa al-Khalifa, was persuaded to reconsider the calling of parliament. After consulting with officials, however, he was reluctant to accept anything beyond the possible establishment of local municipal councils.

Following the outbreak of the war, agitation in religious circles and violent acts increased in almost all Arab Gulf countries. In Kuwayt, it was reported that air raids hit al-ʿAbdali Post (a base on the border line) twice, on November 12 and November 16, 1980. Kuwayt protested, but it did not press the matter further. The government noted the increasing smuggling of arms and explosives into the country and the increasing circulation of political literature critical of the regime. These and other matters were taken as evidence of the existence of a group (or groups) working against the regime. Laws were thus enacted to empower the authorities to limit political activities. Since it became evident that a group of Shiʿi followers, in particular, was very active, ʿAbbas Muhri, the Shiʿi leader, was arrested, stripped of citizenship, and deported with eighteen members of his family. Apart from official gatherings and mosque services, the assembly of more than twenty people without authorization was prohibited. Speeches intended to undermine the authorities were not allowed, and the police were instructed to disperse all gatherings of people suspected of subversive activities. Meanwhile, the police force was reorganized and developed into a more effective instrument to enforce public order. The press, which had enjoyed relative freedom of expression, was brought under rigid control.

On October 5, 1980, following the outbreak of hostilities, the Kuwayti parliament enacted a general mobilization law. Apart from one in Iraq, this law was the first enacted in the Arab Gulf countries. The government was authorized to enforce the law if Kuwayt was either at war or confronted with the threat of war. The High Defence Council, empowered to call up all male citizens between the ages of eighteen and fifty, was entrusted with the enforcement of the law.[9] Moreover, the government began to restrict the entry of aliens into the country and to deport others who had violated the laws of residence and employment. Although Kuwayt had taken a neutral stand in the war, its government began to discuss bilateral security measures with

Iraq. Hoping to influence Kuwayt to distance itself from Iraq, Sadiq Qutb-Zada, Foreign Minister of Iran, had visited Kuwayt in April 1980, but he seems to have failed to alter Kuwayt's leaning toward Iraq.[10]

In Saudi Arabia, Iranian activities were confined to stirring Shi'i followers and spreading propaganda against the regime during the pilgrimage. Following the Iranian Revolution, unrest in the eastern province of al-Hasa, where a considerable number of Shi'i followers reside, was instigated by events in Iran. On November 28, the day of the annual mourning for the martyrdom of Husayn, called the 'Ashura (Muharram 10, 1400), the procession suddenly turned into a demonstration, leading to clashes with the security police. The demonstrators, demanding support for the Iranian Revolution, called on the authorities to stop supplying oil to the United States. They also demanded a more equitable distribution of wealth and an end to discrimination. Very soon, the disturbances spread to other eastern localities, but the authorities quickly dispatched troops to deal with the situation. They also acted with prudence, as some measures to improve the social and economic conditions in the eastern provinces were undertaken, and strict security orders were issued to prevent the recurrence of disturbances. Instigations from Iran, however, continued unabated. Early in 1980, propaganda literature was circulated in which the Islamic Revolution in Iran was extolled. It called on Shi'i followers to criticize the Saudi regime and demand more legitimate reform in the fields of education and culture.

In western Arabia, Iranian instigations were carried out by pilgrims. On September 23, 1981, the Saudi authorities deported a group of eighty Iranian pilgrims from Madina on the grounds that they had violated orders prohibiting political activities during the pilgrimage by distributing pictures of Khumayni. On the following day, twenty Iranian pilgrims were injured in a clash with the security police in Makka; the violence erupted when they (including an Iranian television crew) were not allowed to enter the country.

On December 16, 1981, both Saudi Arabia and Bahrayn announced the arrest of sixty-five Arabs (fifty-two in Bahrayn and thirteen in Saudi Arabia) for their alleged attempt to overthrow the regimes in both countries. Bahrayn declared that the men under arrest confessed that they had received their training in subversive activities in Iran. It was not clear, however, whether those arrested had come from Iraq, as some admitted that they belonged to the Da'wa party, while others insisted that they belonged to the Islamic Front for the Liberation of Bahrayn. On December 18, the government of Bahrayn recalled its ambassador to Iran to protest the alleged coup against it. It also requested that the Iranian ambassador to Bahrayn be replaced presumably because he was in contact with subversive groups. On February 28, 1982, the prosecutor's office in Bahrayn announced that seventy-three persons had been implicated in the abortive coup of the previous December. Thus Shiʻi followers in Bahrayn, acting as a fifth column, seem to have been prompted to seize power from the Sunni leadership and establish an Islamic republic similar to that regime in Iran.

Although the Iranian-inspired disturbances in Arabia have subsided since 1981, they have not completely stopped. Indeed, ever since the Islamic Revolution in 1979, Iran has used the annual pilgrimage as an occasion to propagate its revolutionary ideas in one form of violence or another. In 1987, when several foreign powers began to escort oil tankers in the Gulf, Iran stepped up its efforts to stir trouble during the pilgrimage, presumably on the grounds that Saudi Arabia and other Arab Gulf countries welcomed Western assistance. In fact, only Kuwayt requested the reflagging of its oil tankers, and not only the United States but also the Soviet Union, with which Iran has entered into economic cooperation, agreed to escort Kuwayti tankers. Other powers—Britain, France, and Italy—sent war vessels to protect their own ships and to sweep mines laid by Iran to obstruct free navigation in Gulf channels.

The disturbances during the pilgrimage in 1987 were on a much wider scale. It has become widely known that the pilgrims who stirred the demonstration, equipped with pistols and portraits of Khumayni, had been handpicked and trained for the role they played in leading the demonstrators. Their clash with the police resulted in the death of over 400 pilgrims. Reportedly, 275 were Iranian; 85, Saudi; and 42 pilgrims of other nationalities. It has also been reported that 649 people were injured. Iran's attempt to inflame religious sentiment throughout the Islamic world and escalate tensions in the Gulf did not result in uprisings or the overthrow of Arab regimes, as the Iranian leadership had expected. On the contrary, the Saudi authorities promptly coped with the situation, demonstrating to all believers in the Islamic lands as well as the outside world their concern about the safety of pilgrims and the security of the Grand Mosque at Makka. But the Iranian regime, in inciting disturbances at the Grand Mosque, was not motivated by religion. "Those who consider the Iranian regime, 'religious,' " as Mazhar Hameed, a Saudi political analyst pointed out,

> "must develop a tortuous logic to explain how the *hajj,* the holy pilgrimage, can be demeaned by political demonstrations. This is not extremism on behalf of religion; it is certainly not "fundamentalism"; it is power politics without scruple. As in the use of masses of children to fight the war with Iraq, as in its resort to terrorism, the Iranian regime is extremist in the sense that it recognized no limits in acting to advance its own political interests.[11]

These and earlier disturbances should remind us that the Islamic Revolution in Iran, although claiming to assert Islamic standards, has not always been true to its declared fundamental principles. In Islamic lands—indeed, as in other lands—religion has often been a rationalization of an urge on the part of one country or another for domination over a region inhabited by people who still honor religious values.

Outside Arabia, the two Arab countries that supported Iraq against Iran were Jordan and Egypt. Jordan, to which we have already referred, set the example for others to lend their support for Iraq; but because of limited resources, Jordan's support was essentially political and economic and could not possibly tip the balance in favor of Iraq. At the outset, Egypt had reason to dissociate itself from the Iraq-Iran War, if not to turn against Iraq. Saddam Husayn had taken the lead at the Arab summit meeting in Baghdad in 1978 in the move against Egypt, which virtually ostracized it from the Arab family. There were, however, other overriding reasons that prompted Anwar al-Sadat to seek an end to Egypt's estrangement from Iraq.

First, with the outbreak of the Iranian Revolution, Sadat at once declared himself opposed to the revolution and offered support to the Shah (he even invited him to Egypt), because Khumayni had denounced his peace initiative with Israel and branded him a traitor to Islam. Second, Sadat's peace initiative, undertaken without prior consultation with Arab leaders, led to the isolation of Egypt from almost all Arab lands. By offering Iraq much-needed ammunition and spare parts, which he had obtained earlier from the Soviet Union, Sadat aimed at reconciling Iraq as a step to repair his relationship with other Arab countries. In my conversation with him in 1981, Sadat mentioned another reason: Egypt's acknowledgment of its debt to Iraq for participation in the war with Israel in October 1973. As a reward for Iraq's assistance, Sadat added, he had issued urgent orders to the army to support Iraq, as Soviet arms shipments had been discontinued when the Soviet Union declared strict neutrality in the Iraq-Iran War. Sadat also intimated that he was prepared to visit Baghdad, Riyad, and other capitals to rally support for Iraq. Later, Egypt provided a labor force and military experts to assist in Iraq's war effort. It is estimated that more than 2 million Egyptians served in Iraq, although the number has been considerably reduced since 1984.[12]

Like Jordan and Egypt, Syria and Libya are not Gulf countries; but as members of the larger Arab family, they were expected to support Iraq. Thus their stand with Iran calls for an explanation. We have already noted that Sadat's initiative to make peace with Israel aroused Arab concern that Egypt's single-handed action would weaken the Arab position vis-à-vis Israel. Syria, considering itself the country most directly threatened by Israel, took the lead in 1977 in the formation of the Arab Steadfast Front. Iraq, Libya, Algeria, South Yaman, and the Palestine Liberation Organization (PLO) joined the front, which aimed to mobilize Arab opinion against Egypt's action. Indeed, Syria and Iraq went so far as to enter into a federal union in order to lead the opposition against the Camp David Agreement. In the wake of failure to achieve Syro-Iraqi unity, Iraq withdrew from the Arab Steadfast and Confrontation Front, to the great disappointment of Syria and Libya.[13]

For this reason, when the war between Iraq and Iran commenced, Syria took a neutral attitude at the outset, claiming that the war was an Iraqi venture to assert its leadership in the Gulf region and a diversion of Arab resources whose primary beneficiary was none other than Israel. It also maintained that Iraq had undermined the Arab position by going to war with a country that was potentially sympathetic to the Arabs against Israel. Iraq, rejecting Syria's allegations, insisted that its war with Iran was in defence of its territorial integrity, instigated by Iranian border attacks and refusal to return the three central land sectors belonging to Iraq. Iraq also reproached Syria for having delivered weapons to Iran in exchange for oil and other commodities and for no longer remaining neutral. It is true that Syria allowed Iraq in the first year of the war to export oil through the trans-desert pipeline to European markets, and a new agreement to the satisfaction of both sides was concluded in 1982. But within three months, the pipeline was completely shut down, and Syria was delivering arms and ammunition to Iran.[14]

Syria's economic difficulties might have contributed to Asad's hostile attitude toward Iraq. Despite Saudi's financial assistance, estimated at $540 million a year, Syria is heavily burdened with debt. It owes Iran over $2 billion for oil. It also owes the World Bank and Western countries some $3.5 billion, and its debt to the Soviet Union for military arms is probably over $15 billion. Heavy military spending has taken most of its revenue. Lack of improvement in domestic economic conditions, despite an increase in oil production—an important discovery in 1984 has increased oil production by one-third or more—has rendered Syria largely dependent on Iranian oil. To keep Syria as an ally against Iraq, Iran in early 1987 seems to have concluded a new oil agreement, by which Syria will be provided with 1 million tons of free oil and 2 million tons at 25 percent less than the OPEC price over a year. Small wonder that Syria is not inclined to sever its relations with Iran unless a major disagreement erupts over Lebanon. As Nora Bustani (Boustany), quoting a high Western diplomat, observed, "Maybe at some point, the costs the Iranian impose on Syria in Lebanon are going to be more important in Syrian eyes than what the Iranians are doing to the Iraqis, but only then will there be a shift in alliances."[15] The possible alternative to Syria's alliance with Iran would be to call to life the Syro-Iraqi unity scheme, as Asad himself seems to have suggested, which would *ipso jure* compel Syria to sever its relations with Iran. Such an arrangement,

giving Asad a free hand to reduce Iran's influence in Lebanon, would end
Syria's isolation in the Arab world and enhance both the Syrian and the
Iraqi positions.

Several Arab countries on good terms with the Syrian and Iraqi regimes
have on more than one occasion tried in vain to bring about a reconciliation.
Apart from official attempts, noted before, Saudi Arabia and Kuwayt have
tried through special emissaries to persuade Asad and Saddam Husayn to
reconcile their differences.[16] During the past two or three years, when the
ruptured relations between Syria and Jordan have been repaired, King Husayn joined the Saudi leadership to try to bring about reconciliation. President Asad was asked to stop supporting Iran, as the Iranian regime had taken
the offensive and invaded Iraqi territory. Asad, claiming that he had been
assured by the Iranian leadership that it had no desire to acquire Arab territory, said that the war was and remains against the Iraqi regime and not
against Iraq per se or other Arab Gulf countries. In a conversation with King
Husayn in 1986, Asad shared Husayn's concern that the Iraq–Iran War had
become no less dangerous to Arabs than the Israeli threats. When, however,
he was asked to stop military assistance to Iran, he maintained that Syria
and Iraq should first unite under one leadership so that both could face Iran
as one state. To this, the Iraqi leaders replied that unity might complicate
the situation and should be discussed after the war ends. At a closed meeting, arranged by King Husayn of Jordan and Crown Prince ʿAbd-Allah of
Saudi Arabia, the presidents of Iraq and Syria finally met to iron out differences on April 26 and 27, 1987, at a desert rendezvous, known as H4 (one
of the pump stations on the old Iraq–Palestine pipeline), across from the
Iraqi border. In their tête-à-tête conversation (most of the time alone), they
reviewed practically all the issues on which they disagreed and came to a
tacit agreement on most of the specific issues, save the two major questions
of unity between the two countries and Syria's stand on Iran's side. Asad
agreed to stop his support of Kurds engaged in military activities against
Iraq. Syria, once a supplier of Soviet arms to Iran, had already stopped
providing any war matériel. The two leaders, long engaged in subversive
activities against each other's regimes, agreed to end these ventures. The
possible resumption of the flow of oil through the Syro-Iraq pipeline, closed
since 1982, was also discussed. In lieu of Iranian oil, Saudi Arabia offered
Syria 50,000 free barrels a day, and presumably the reopening of the closed
Iraq pipeline might meet part or all of Syria's oil requirements. It was agreed
that the details concerning bilateral issues would be dealt with at forthcoming meetings between the prime ministers and the Ministers of Interior and
Oil of the two countries. Asad's reservations about his country's attitude
toward the war must have undergone some change, as he seems to have

agreed to postpone discussion about the possible unity of Syria and Iraq, until the war ends. Finally, it was agreed that once all the details were dealt with by the top ministers of the two countries, a joint declaration about normalization of relations would be declared after the two leaders had met again later in the year. When Asad and Saddam Husayn met again during the Arab summit meeting in Amman (November 1987), Asad finally agreed to join in the Arab leaders' condemnation of Iran for its refusal to accept the United Nations Resolution 598, which called for a cease-fire and withdrawal of forces to international frontiers. This step may well lead eventually to the end of the antagonism, if not to the full normalization of relations between the two Ba'thist regimes in Syria and Iraq.[17]

The Syro-Iraqi rivalry has not only undermined Iraq's position in the Gulf War, but also directly contributed to the prolongation of the war, as it prompted Libya, a close ally of Syria, to send missiles and other matériel to Iran. It also encouraged Syrian Kurds to assist Kurdish factions in Iraq and Iran who participated in military operations in the northeastern sector of Iraq. These activities have been looked on with suspicion and disfavor in Pan-Arab circles and considered by Iraqi nationalists as a betrayal of Arab nationalism, contrary to Syria's claims that it sought to promote Arab interests and Pan-Arab ideals. For these and other reasons, Jordan and Saudi Arabia have been spurred to bring about reconciliation between the Syrian and Iraqi leaders, hoping that it would contribute to a final settlement of the war.[18]

Libya's attitude toward the Gulf War was determined by several factors— partly because of its commitment to Syria, with which it had forged a political union, and partly because of its conflict with the United States, leading indirectly to quarrels with moderate Arab states, such as Saudi Arabia and Jordan. After the outbreak of the Iraq-Iran War and the dispatch of the American AWACS (Airborne Warning and Control System) aircraft to Saudi Arabia, Colonel Qadhdhafi, in a speech delivered on October 19, 1980, on the occasion of 'Id al-Adha (Feast of the Sacrifice), referred to the presence of the AWACS in Saudi Arabia as "occupation," and he called for a jihad to free the holy sanctuaries from this contamination.

Last but not least is Qadhdhafi's involvement in the disappearance of Musa al-Sadr, the Shi'i spiritual leader in Lebanon, who visited Libya in August 1978, at the invitation of Colonel Qadhdhafi, and suddenly disappeared. Since Sadr was a protégé of the Ayat-Allah Khumayni, who suspected that Qadhdhafi was directly responsible for Sadr's disappearance, Libya's relationship with Iran was bound to be affected by the Sadr affair. Apart from Libya's Pan-Arab policy, which would have been denounced as secular and antireligious by Iranian leaders, Qadhdhafi was concerned about Khumayni's possible call for an investigation of Sadr's disappearance in Libya were he to take Iraq's side in its conflict with Iran. To avoid Khumayni's possible denunciation of his policy, Qadhdhafi declared his support of the Iranian Revolution, and offered his country's economic cooperation and assistance in weaponry. At first, the Iranian regime refused Qadhdhafi's offer of establishing diplomatic relations, but the Iranian leaders were in need of Libya's military assistance and persuaded Khumayni to drop his inquiry about the fate of Sadr. It took over two years before Khumayni finally agreed to send an ambassador to Libya. Despite his support of Iran, Qadhdhafi has not yet been able to visit Iran, since such a visit would arouse the sensitivity of the Shi'i community in Lebanon as well as the friends and supporters of the Sadr family in Iran. In justifying his support of Iran against Iraq, Qadhdhafi said that Iran, as an Islamic country, is entitled to a greater Libyan support than Iraq, even though Iraq is a member of the Arab family.[19]

Qadhdhafi's vacillation between Pan-Arabism and Pan-Islamism and the subordination of Libya's foreign policy to his personal whims has alienated his country from most Arab lands. Even Syria's unity with Libya, an artificial arrangement with Asad, may be broken at any moment, just as other unity arrangements with Tunisia and Morocco have been broken. When asked by Iraqi leaders why he had been supporting Iran against Iraq, he replied that he was not against Iraq but merely tried to maintain friendly relations with Iran. Nonetheless, he had supplied Soviet missiles and ammunition, which Iran used against Iraq. However, since Syria stopped sending military equipment to Iran, Qadhdhafi has also stopped his military support. Unlike Syria, his support of Iran was not because of Libya's need for oil. For this reason, Iraq—indeed, several other countries, such as Saudi Arabia (each for its own reason)—severed diplomatic relations with Libya in 1981, and Libya's relations with other countries in Asia and Africa are strained. Qadhdhafi has maintained friendly relations with only socialist countries, from some of which he has purchased weapons.[20]

Although neither an Arab nor a Gulf country, Turkey is a neighbor of both Iraq and Iran and could hardly be expected to remain indifferent to the events that have been taking place across its borders since 1980. Turkey's relations with its two neighbors are deeply rooted in history. For almost four centuries, the Ottoman and Persian armies often fought many battles on Iraqi territory; Persia sought to control the Shi'i sanctuaries and perhaps the entire Shi'i community in Iraq, which the Ottoman sultans were determined to perpetuate their domination over the entire Iraqi territory. Following the First World War, the new Turkey lost control over Iraq (as well as over other Arab lands) and was no longer involved in the confessional and frontier problems that Iraq had inherited from the Ottoman Empire. It made an attempt to preserve its control over the province of Mawsil (Mosul), in which the Kirkuk and Sulaymaniya districts (inhabited, in the main, by Kurds and Turks) were included; but it lost its claim to that province when its dispute with Iraq was referred to the League of Nations in 1926.[21] Rich in oil and important for its strategic location, Iraq could hardly form a viable state were it to be reduced to the provinces of Baghdad and Basra. Since a large portion of the Kurdish community in the Mawsil Province remained within Iraqi territory, Turkey and Iraq agreed to consult with each other should the Kurdish activities in one country create repercussions in the Kurdish community of the other.

When the Iraq-Iran War broke out, Turkey declared its strict neutrality. Whether at the United Nations or at other international councils, it voiced its support for all attempts to restore peace between the two neighbors. It attended the third Islamic summit, held at Makka (in 1981), and participated in the commission set up to mediate an end to the war.[22] But Turkey could not completely extricate itself from the war when the Kurdish community became involved in it. Two or three years after the war began, Kurdish activities instigated by Iran had no direct impact on the Kurds in Turkey. But in 1983 and 1984, Kurdish nationalists in Turkey, already organized as an underground Marxist group called the Kurdish Workers' Party (PKK), made several incursions across the Iraqi border into an area (varying between ten and fifteen miles in depth) extending from the Iranian to the Turkish border that the Iraqi government had declared to be a security zone and from which it had transferred the Kurdish inhabitants to nearby areas in Iraq, generously compensating them for the farms they had abandoned in the security zone.[23] The security zone, rid of restless leaders who fled the country to Iran, remained quiet until the war broke out in 1980. Since Mulla Mustafa had died (March 1, 1979) over a year before the war started, his son Mas'ud and some of his followers, encouraged by Iran and Syria to undermine the Iraqi regime, began to infiltrate into the security zone. Despite the presence of an Iraqi force in the area to stop the infiltrations from Iran, the incursions continued unabated, and the clashes between Kurdish factions and the Iraqi army were often reactivated whenever the Iranian authorities dispatched reinforcements.

These events coincided with the renewed activities of the PKK in Turkey. At first, the PKK incursions were unrelated to Kurdish activities in the Iraqi security zone. When some of the PKK rebels escaped into the security zone, the Turkish authorities sought to pursue them. Seeking support from Kurds infiltrating into the security zone from Iran, the PKK rebels began to cooperate in activities against both Turkey and Iraq. It was, of course, in the interests of both Iraq and Turkey to cooperate in stopping these incursions into the security zone. To achieve this end, Turkey and Iraq concluded an agreement early in 1984 for one year (it would lapse if not renewed each year) by which Turkish forces would be allowed to cross the Iraqi border to arrest the Turkish (Kurdish) rebels in hot pursuit, provided that the Iraqi authorities were notified and the Turkish forces withdrew immediately after each operation was completed. Moreover, Turkey was allowed to shell or bomb centers in the zone where the rebels were suspected to have taken refuge.

The most important incursion since the agreement came into force took place in 1987. Early in March, an Iranian force penetrated beyond the security zone and took control of Mount Kardamd, overlooking the town of Haj Umran. Since this town is about sixty miles from Kirkuk, where the Iraq–Turkish pipeline begins, it was suspected that the Iranian force might advance beyond that area to stop the flow of oil from Kirkuk to Turkey (although this was unlikely, owing to the rugged mountain ranges). Nevertheless, Turkey warned the Iranian regime against such an action, as it was considered an interference in Turkish domestic affairs. Turkish suspicion was confirmed when the Iranian press published unfavorable statements made by Iranian leaders in which the Turkish regime was denounced as antireligious. Since Turkish (Kurdish) rebels had been infiltrating into the security zone, Turkish aircraft bombed several centers to which the rebels had fled, on the grounds that the Turkish Kurds had come into a clash with local Kurds (presumably Iraqi Kurds returning from Iran). The Iranian regime protested Turkish violations of international frontiers, although the Turkish authorities had notified Iraq about the action, in accordance with the Turko-Iraqi agreement. Meanwhile, the Iraqi government had dispatched a force to the Haj ʿUmran town; it drove out the Iranian forces in a costly battle to both sides. But these Iranian-Kurdish incursions continued to recur, especially in the area near western Azarbayjan, and the Iranian authorities, in cooperation with Kurdish factions opposed to the Iraqi regime (for example, the Talabani and Masʿud groups), have found them useful for diversionary military tactics in their struggle with Iraq.[24]

With regard to its economic relations with Iraq and Iran, Turkey benefited much from the war. At first, the Iraqi (Kirkuk–Alexandretta) pipeline was damaged and remained closed until the end of 1980. By an understanding between the two countries, the pipeline was repaired. Turkey, in control of water dams, promised that the Euphrates waters would be increased to meet Iraq's need for irrigation for an eight-year period of development in southern Iraq. Although no full commitment about the increase of water was made to Iraq, the capacity of the pipeline (650,000 barrels a day) was increased by 500,000 barrels a day to over 1 million barrels a day, as Turkey had become the principal terminal for Iraqi oil exports. In an agreement signed in 1985, another pipeline, to be completed in two years, would double the amount of oil exported. Perhaps no less important, Turkey has become an increasingly important source of foodstuffs for Iraq. Iraqi and Turkish delegations exchanged visits almost every year to facilitate the export of Turkish products as well as to arrange to build bridges over the Euphrates and to improve highways for the transport of commodities between the two countries. Evidence of this increased commercial relationship can easily be observed on the newly reconstructed highways between Iraq and Turkey through the northern provinces.[25]

With regard to Iran, Turkey sought to improve its economic relationship by entering into several trade agreements to import oil for its domestic requirements and to export foodstuffs and other Turkish products. One such agreement dealt with the import of material for the improvement of railways, highways, and other transport lines. Thus Turkey's bilateral economic agreements with Iraq and Iran proved an asset to both sides, although it was not unnatural that technical problems often hampered the full implementation of these agreements.

Turkey's experiments in bilateral trade agreements with Iraq and Iran are likely to continue after the war, as Turkey will always be in need of oil and gas for domestic requirements. Iraq and Iran may also continue to import foodstuffs, although as agricultural countries, their need for farm products in peacetime may not be as pressing as in wartime. Iraq's need for irrigation water, however, is likely to increase after the war, and the bilateral trade relationships might even become more important for the mutual benefit of both countries. Peace and security in the region as a whole is in Turkey's interest, and its improving relationships with Saudi Arabia and other Gulf countries are likely to expand trade relationships, especially with countries in need of agricultural products.

# XI

# In Search of Peace and Security in the Gulf

Bordered by oil-rich states, the Gulf has become the focus of world attention, and the question of peace and order is of concern not only to the Gulf countries, but to almost all others that have an interest in the region as a whole. As the war between Iraq and Iran has demonstrated, any conflict between one Gulf country and another, whether inside or outside the Gulf, may hurt the interests of all concerned and invite foreign intervention.

Why have the Gulf countries, unilaterally or collectively, been unable to maintain peace and order without foreign intervention?

Some have answered the question simply by stressing rivalry among Gulf leaders that has impeded the establishment of an effective collective-security system to deal with regional security problems. Nor have the superpowers, owing to their conflicting interests, been able or willing to exercise control over their clients and friends to prevent regional conflicts. The situation is compounded by intricate traditions and the legacy of colonial domination, which have not yet been superseded by modern patterns of authority and diplomatic procedures. All these forces and conflicting interests have contributed in no small measure to the prevailing tension and disorder in the Gulf.

From early modern times, the region was dominated by two Muslim powers, the Ottoman and the Persian; the first controlled the area extending westward from the head of the Gulf to the Mediterranean, and the other dominated the whole eastern Gulf area. Rivalry and conflict between these two determined adversaries were not always confined to the northern area of the Gulf; they often extended southward, and the whole region fell into disorder and chaos. Small wonder that the Gulf had become the theater of active privateering and was torn by local dynastic rivalries that invited foreign intervention when both the Ottoman and Persian Empires became weak and exhausted. British entry into the Gulf in the early nineteenth century, at first in pursuit of trade and later to protect the Indian subcontinent, was justified on the grounds of disorder and the prevalence of piracy, although the underlying motive for piracy was local resentment of foreign intervention and not necessarily the desire for personal gain.[1]

For over a century and a half, Great Britain played the predominant role in the maintenance of peace and order at a relatively small cost by inviting the rulers of the Gulf principalities (often called shaykhdoms) to stop privateering and making war with one another and to enter into a series of bilateral agreements that would prohibit privateering and warfare. Gradually, almost all the Arab principalities and Persian coastal provinces passed under one form of British control or another before they finally regained their full independence, some after the First World War and others after the Second World War.[2]

Britain's decision in 1969 to withdraw its military presence from the Gulf by the end of 1971 led to a tacit understanding among the major powers that responsibility for the maintenance of peace and order should fall on the Gulf states themselves. From the regional vantage point, there were two possible approaches to the task that Britain had fulfilled in the past.

First, one or two of the major Gulf states—Iran or Saudi Arabia—would assume security responsibility either singly or jointly. This arrangement, in line with British imperial traditions, may be called the task of the policeman.[3]

Second, all the states, on both sides of the Gulf, would assume collective responsibility by creating a regional organization entrusted with security matters. This arrangement, comparable with that of the Organization of American States (OAS), may be called, in accordance with the United Nations Charter, the regional-security arrangement.

Since the whole eastern Gulf coast was under Iran's control and Saudi Arabia had not yet become militarily equipped for the role of policeman, the obvious candidate was Iran. The Shah of Iran, who had long been impressing on the United States the need for security, seized the opportunity of British withdrawal to assert his claim to Gulf leadership as a means to check possible Soviet penetration into the Gulf, although the United States would have preferred a joint leadership (Saudi and Iranian), which American policy makers often referred to as the twin pillars. The Shah's ambition to play the role of policeman, however, alarmed some of the Arab Gulf states, especially Bahrayn, to which Iran had claimed sovereignty. To assure the Arab states that he had no territorial ambitions, he recognized Bahrayn's independence following the termination of British protection in 1969, thereby tacitly renouncing Iran's territorial claim to these islands. Contrary to his assurances, however, the Shah annexed the islands of Abu Musa and the two Tunbs, belonging to the United Arab Amirates, which naturally aroused the suspicion of the lower Arab Gulf states.[4]

The Shah's dependence on American endorsement of his rearmament program, although enhancing his country's position in the Gulf, aggravated opposition to his regime internally, since it implied American approval of his mode of governance. After the Shah's fall and the almost complete disarray in the military, the new Iranian regime could hardly be considered capable of exercising security responsibilities. Nor was it trusted by its neighbors because of its outspoken religious fanaticism and its declared intention to export the Islamic revolution to other countries.

In these circumstances, Iraq might have seized the opportunity to undertake a unilateral or joint Arab Gulf security responsibility, in accordance with its Arab National Declaration of 1979. But it chose to settle its accounts with Iran before becoming involved in Gulf affairs. Nor did its alliance with the Soviet Union, concluded in 1972, although primarily to obtain arms, prove helpful when it wanted to move closer to the West and other Gulf countries. Only after it had gone to war with Iran did Iraq's relationships with the West and the Arab Gulf countries begin to improve.

The Iraq-Iran War prompted the Arab Gulf countries to assume larger measures of responsibility through collective-security arrangements as an alternative to unilateral initiatives. Since the Shah's role as the Gulf policeman, disguising an Iranian hegemonic posture, was resented by the leaders of other Gulf countries, although none was then ready to challenge him, the collective-security system seemed the natural substitute, and it received the encouragement of Western powers. Nor was such a collective-security system unprecedented in the Arab world. The League of Arab States was recognized by the United States as a regional-security arrangement in whose membership the Arab Gulf states were included. Its pact, supplemented by the Treaty of Joint Defence and Economic Cooperation, provided an adequate instrument for an Arab collective-security system. Why did the Arab Gulf states, it may be asked, not seek recourse to the Arab League for Gulf security?

Like other regional-security systems, the Arab League was established in response to certain immediate requirements. Following the Second World War, the Arab countries were preoccupied with two fundamental issues: first, to protect Arab interests against Zionist claims to establish a national home in Palestine, and second, to achieve unity and protect Arab interests, although the form of Arab unity and the means to achieve it were still vague and undefined. But when Israel was established in 1948, the Arab League was unable to stand up to its responsibilities.[5] Apart from promoting economic and cultural cooperation, the task of the Arab League was thus re-

duced in practice to defending the Arab countries from Israel's encroachments.[6]

It was for this reason that the pact of the Arab League, leaving much to
be desired, was supplemented in 1950 by the Treaty of Joint Defence and
Economic Cooperation to provide security measures for Arab League members. Since this instrument was drawn up before most of the Arab Gulf
countries became independent, it did not take into account their security
requirements. Thus when the Arab Gulf leaders met to discuss Gulf security
and economic cooperation, it was not unnatural that they looked into other
means of security relevant to challenges emanating from conditions in the
Gulf.

An inquiry into Gulf security, however, should not be viewed from only
a regional-security perspective; it should also be perceived from the global-
security requirements of the superpowers, to whose area of rivalry and conflicting interests the Gulf countries have been drawn. During the Second
World War, the Soviet Union seems to have aspired to extend its influence
or control over an area extending from the Caucasus to the Indian Ocean, a
legacy of imperial Russia's traditional policy of acquiring warm-water out-
lets. In their negotiations with Nazi Germany in 1940, the Soviet leaders
revealed their long-cherished aspirations to extend their control over the area
(in Molotov's words) "in the direction of the Persian Gulf and the Arabian
Sea."[7] The Soviet Union not only expressed its demands in official conversations, but also sought to realize them by action when its forces took control of northwestern Iran (Azirbayjan) and encouraged Kurdish nationalists
in the establishment of the Republic of Mahabad in 1946. After the Second
World War, when the Soviet forces failed to withdraw, contrary to the Tihran Declaration (1943), in which Britain, the United States, and the Soviet
Union pledged to withdraw their military forces, the Soviet Union was forced
to withdraw from Iran by an action of the United Nations.[8]

The Truman Doctrine (1946), promising support to Greece and Turkey
against Communist penetration, was the American reply to renewed Soviet
interest in the Middle East. But the Truman Doctrine did not specifically
include the Gulf region, whose defence remained Britain's responsibility for
another quarter of a century. When Britain finally withdrew its military presence from the Gulf in 1971, the Shah of Iran took upon himself the responsibility for the security of the Gulf. With American blessing, the Shah played
his short-lived imperial role until his fall in 1979.[9]

Two events seem to have dramatically altered American policy toward the
Gulf region. First, the unexpected outbreak of the Iranian Revolution in
February 1979, and, second, the Soviet invasion of Afghanistan later that
year. The two events were not unrelated, and they revealed once again that

Gulf security could not be completely divorced from global security, although the impact of the Soviet invasion of Afghanistan on American policy makers seem to have been more profound than the Iranian Revolution. There was a difference of opinion among President Jimmy Carter's advisers about the impact of the two events of Gulf security before Carter finally announced his policy toward the Gulf. Cyrus Vance, Carter's Secretary of State, saw in the Soviet invasion of Afghanistan a more limited objective than the ultimate Soviet goal to reach the Indian Ocean and the Gulf. The immediate objective, Vance said, was primarily to "protect Soviet political interests in Afghanistan which they saw endangered . . . by a spread of 'Khomeini fever' to other nations along Russia's southern border," although he realized that the global impact might present threats to American interests in the Gulf region.[10] Vance suggested caution, but Carter, prompted by long-term global-policy considerations (stressed in particular by Secretary of Defense Harold Brown and National Security Adviser Zbignew Brzezinski), proceeded to announce what came to be known as the Carter Doctrine (specifically in reference to Gulf security), intended as a warning to Soviet global pressures. Since Leonid Brezhnev himself had ordered the Soviet invasion of Afghanistan, Jimmy Carter quickly responded to counteract possible Soviet penetration into the larger region between Afghanistan and the Gulf. In the light of subsequent developments, both Brezhnev and Carter seem to have overreacted, although each was spurred by an entirely different perspective.

The Soviets, in invading Afghanistan, may be regarded as having disregarded the lessons of history. Both Russia and Britain, after protracted wars to control Afghanistan known as the Afghan wars (especially in 1838 and 1878), came to the conclusion that it was not in their best interests to meddle in Afghan affairs. They felt compelled to recognize the country's neutrality and kept out of domestic affairs, as provided under an Anglo-Russian accord (1880).[11] Carter wisely decided to leave the task of resisting the Soviet invasion of Afghanistan to its people, known for their valor and resilience against foreign invasions, and offered only support with weaponry through neighbors; but Brezhnev's successors, although realizing the futility of their action, have not yet been able to extract themselves from the dilemma. With regard to the Iranian Revolution, Carter's concern about the safety of American hostages in Tihran seems to have distracted his attention from the revolution and discounted its long-term impact on Gulf security. He thus was more concerned about the immediate impact of the Soviet invasion of Afghanistan on the Gulf than about the long-term impact of the Iranian Revolution.

Carter made his warning to the Soviets to keep out of the Gulf quite clear. In his State of the Union address to a joint session of Congress (January 23,

1980), he announced what came to be known as the Carter Doctrine. He declared, "Any attempt by any outside force to gain control of the Persian Gulf region will be regarded as an assault on the vital interests of the United States of America and such an assault will be repelled by any means necessary, including military force."

"The Carter Doctrine," wrote Brzezinski, "was modeled on the Truman Doctrine, enunciated in response to the Soviet threat to Greece and Turkey. . . . The collapse of Iran [under the Shah], and the growing vulnerability of Saudi Arabia dictated the need for such a wide strategic response." [12] Brzezinski maintained that American security in the broad sense of the term had become "interdependent" on the security of other regions, in particular Western Europe, the Far East, the Middle East, and the Gulf region. It is for this reason that "a regional security framework" without a formal alliance would be necessary as a means to provide the needed military support against Soviet pressures and possible intervention. For the implementation of military support, he proposed that naval and air-base facilities be extended by some of the Gulf countries to establish an American military presence in the area. [13] During 1980, the United States had acquired access to Masira, an island close to the Gulf belonging to ʿUman, and supporting bases in Barbara (Somalia), Mombasa (Kenya), and Raʾs Banas (Egypt)— somewhat away from the Gulf—but none was extended by Gulf countries, save ʿUman.

Brzezinski's vague concept of a "security framework" may mean either that the responsibility of defence would depend on the Gulf countries themselves, supported by American military force, to which Carter referred in his State of the Union address, or that the extended regional facilities would eventually become military bases under American control. [14] After he left office, Brzezinski said that he had preferred the latter alternative and not merely the "cooperative security framework" referred to in Carter's State of the Union address. "With the earlier British disengagement from 'east of Suez' [1971]," Brzezinski added, and "with the collapse of our strategic pivot north of the Persian Gulf, I felt that a wider response by the United States was needed." [15] After the outbreak of the Iraq-Iran War, Brzezinski's alternative of an American military presence seemed to both Carter and Harold Brown, Carter's Defense Secretary, more relevant to altered conditions in the Gulf. Serious discussions about supplying weapons to some Gulf countries (especially Saudi Arabia) and stationing American forces in the region took place in the National Security Council. While Saudi Arabia welcomed the idea of "cooperative security," it accepted it not in the form of an American base on Saudi soil, but as a Saudi military buildup with American weaponry and professional expertise that it paid for. Apart from ʿUman, no Arab Gulf country was prepared to allow American military forces to be stationed on its territory in peacetime, as the question of the presence of foreign forces on Arab territory following the withdrawal of British forces has become a sensitive issue in Arab domestic politics.

Shortly after the start of the Iraq-Iran War, four American AWACS planes and other military equipment were stationed in Saudi Arabia, in accordance with the proposal to establish cooperative security arrangements. Saudi Arabia had already asked for the purchase of the planes, and Carter approved the Saudi request on September 28, 1980. The deployment of the AWACS and the professional personnel service were taken as evidence of American intention to support Arab Gulf countries against possible foreign intervention.

American strategy for Gulf security, although criticized in congressional hearings, was endorsed by the Reagan administration when it succeeded Carter in January 1981.[16] Indeed, at the outset, the Reagan administration seems to have contemplated an even more ambitious military plan for Gulf security. The concept of the Rapid Deployment Force, although discussed under the Carter administration, had not yet been adequately implemented. The Reagan administration sought to extend it. It regarded the absence of an American military presence as a tacit invitation to Soviet intervention. To deter Soviet initiatives, Alexander Haig, Reagan's Secretary of State, visited the Middle East in April 1981, and proposed to enter into bilateral arrangements with Gulf and other neighboring countries that would formally permit American forces to be stationed on their territories. Such arrangements, called "strategic consensus," aimed directly at deterring Soviet penetration.[17] As a direct reply to Soviet regional activities, weapons were extended to countries that bordered countries that had passed under Soviet influence or control, such as Afghanistan, South Yaman, and Ethiopia. In the implementation of this policy, the United States sought to rely on its own military force whenever there would be need to deter Soviet intervention, but the only concrete action that emerged was the stationing of naval power in areas near the Gulf. Nor were the funds to maintain large enough forces ($5.5 billion in 1981 and 1982) adequate.[18] American allies in Europe were invited to participate in the Indian "proximity force," but only Britain and France responded with modest contributions (Britain dispatched six warships and France five when the Gulf was threatened to be closed to Western shipping). Above all, the reluctance of the Gulf countries to permit American forces on their territories proved the greatest stumbling block. Sensitivity to the existence of foreign "bases" frustrated formal military commitments. Even Egypt, which promised the use of Ra's Banas as "facilities" at times of crisis and not as "bases," was criticized by other Arab countries, although the United States promised to give planes and weapons to the Egyptian army. The assassination of Sadat was a reminder that public opinion was not in favor of his arrangements with the United States.

With regard to Saudi Arabia, the major oil country in the Gulf region, the Carter administration tried to persuade the Saudi government to allow the stationing of military forces on its territory not only to defend against aggression, but also to deter Soviet intervention in the Gulf. The Reagan administration seems to have tried again to persuade one or more of the Gulf countries to accept some form of military presence on their soil, but none save 'Uman feel the need for it, although all are prepared to accept military cooperation in principle. In the Middle East, no sovereign state would normally permit the stationing of foreign forces on its territory unless it felt there was an imminent threat of foreign aggression. Neither Saudi Arabia nor other Gulf states felt that they were under Soviet threat or aggression to warrant the existence of American forces on their soil. If there were ever a threat, they maintained, it would be from Israel and not from the Soviet Union. Indeed, the existence of foreign forces or bases on Arab territory had become a symbol of imperialism. Saudi Arabia, which has not experienced foreign control—at any rate, not in the form of European domination—was not expected to yield to American demands that would compromise its sovereignty in the eyes of both its own people and Arabs in other countries.[19]

The Reagan administration tried to persuade the Saudi leaders that the American forces would not only deter Soviet intervention, but also become a shield for the regime against possible internal upheavals. The press, echoing critics of the Saudi leadership, began to publish articles describing the regime as unstable and liable to be overthrown at any moment. Some writers went so far as to argue that in the event of an internal uprising or foreign intervention, American forces should occupy the area where the oil fields exist before they fall into the hands of foreign forces. Others, in defence of the Saudi regime, stressed the religious and historic background of the dynasty and the traditional friendship with the West.[20]

The outbreak of the Iraq–Iran War provided the occasion of "military cooperation" between Saudi Arabia and the United States. On September 29, 1980, four American AWACS aircraft with air crews and ground personnel were dispatched to Saudi Arabia as a necessary military precaution to deter possible foreign intervention. In November 1980, the Carter administration approved the sale to Saudi Arabia of the planes and arms it had requested, including the AWACS aircraft that had already been dispatched. Under the Reagan administration, Saudi Arabia renewed its request for arms and asked for additional arms (\$8.5 billion) to defend itself against possible attacks by Iran. The next step was that Congress had to be notified before the AWACS were delivered or the other military requests were met.

Even before Congress was officially notified, a campaign in the press depicted the Saudi regime as suffering from inherent instability and claimed that the arms would be a shield for survival. If the regime were incapable of surviving, it was argued, the American arms would fall into the hands of a hostile regime, just as American arms in Iran fell into the hands of the Khumayni regime. The Saudi government warned that if it could not buy arms from the United States, it would purchase them from England or Germany, and it insisted that the arms deal would be the test of American friendship.[21]

On September 9, 1981, Reagan sent Congress an informal notification of intent to sell the arms to Saudi Arabia, with the formal notification to take place on October 1. The vote was expected to take place within thirty days. Discussion of the domestic aspect of the controversy is deemed outside the scope of this study. Suffice it to say that the objections raised both inside and outside Congress may be summed up in four fundamental points. First, the sale of arms to Saudi Arabia was considered a major threat to Israel's security. Second, Saudi Arabia was not really threatened by neighbors, nor were the arms adequate to deter Soviet threats. Third, the sale would put some of the most sophisticated American aircraft and equipment at risk of Soviet acquisition. Fourth, Saudi Arabia opposed American efforts to advance Arab-Israeli peace negotiations. The argument in favor of the sale stressed Saudi security requirements not only for its own defence, but also for the security of the Gulf as a whole against foreign aggression. The sale was also regarded as a test of friendship toward Saudi Arabia. Congressional rejection, it was argued, would have a ''profound negative impact'' on the future of American-Saudi relations.[22]

When congressional opposition seemed to be gaining strength, Reagan took up the matter himself. Reproaching pro-Israeli critics, he said, ''It is not the business of any other nation to make American foreign policy.''[23] On October 29, 1981, the Senate approved the sale by a majority of 52 to 48. It was a victory for the policy of cooperation with Saudi Arabia on Gulf security. In reply to critics who warned about Saudi instability, he declared, ''We will not permit [it] to become another Iran.''[24] In 1986, when the Saudi government requested further arms purchases, and Reagan's conversations with congressional leaders were not very encouraging, the Saudi government decided to diversify its weaponry by purchasing Tornados from Britain without restrictions. While this arrangement relieved Reagan of further confrontations with Congress, it did not adversely affect the Saudi special relationship with the United States, as further American missiles were made available.[25]

To the Soviet Union, Carter's and Reagan's policy toward the Gulf had already become clear: to enter into a set of bilateral alliances with Arab nations, irrespective of local differences, against Soviet penetration, real or fancied. This policy, in Soviet eyes, was regarded as part of the larger American design to counterbalance the growing Soviet influence in the Third World. The American thrust into the Gulf region may be regarded as an extension of American influence into areas where Soviet influence had been shrinking, not to speak of the region from the Arabian Sea to the Indian Ocean, where NATO naval power already prevailed.

In an effort to restore its prestige and influence in the Middle East, the Soviet Union called for an international agreement to neutralize the Gulf. This proposal, first made by Brezhnev in India in December 1980, was reiterated in 1981 as a reply to the Carter Doctrine and Reagan's "strategic consensus" for the Gulf security. Brezhnev also reiterated his call for the termination of the Iraq-Iran War, concerning which both the United States and the Soviet Union had declared their neutrality, and the settlement of the Arab-Israeli conflict by negotiations under the auspices of an international conference in which the Arab states (including the PLO), Israel, the United States, the Soviet Union, and some European powers would participate. These proposals, appealing to such moderate Arab countries as Saudi Arabia, Kuwayt, and Jordan, which had already voiced similar views, were intended to enhance Soviet prestige in the Middle East.

But Soviet policy toward the Gulf was not expressed merely in broad declarations and pious statements. Just as the United States sought by private and formal conversations to discuss specific issues, so the Soviet Union initiated contacts with almost all Gulf countries and other interested parties in the region. In Soviet talks with Syria, Libya, and Algeria—members of the Arab Steadfast and Confrontation Front—there was an agreement on the Soviet proposal that the Gulf and the Mediterranean Sea become "zones of peace" immune to foreign intervention. Since Algeria and Libya are far from the Gulf and Libya's main cause of concern was the United States, Syria became the most convenient conduit for the occasional transfer of Soviet arms to Iran. Syria may or may not have been encouraged to play this role, but its bitter conflict with Iraqi leaders seems to have been the primary reason for its action. Although the Soviet Union is the friend and close ally of Syria, Libya, and South Yaman, it did not enjoy the same degree of prestige in Arab Gulf countries, despite its having established diplomatic relations with some of them. The Arab Gulf countries are jealous about their newly won independence, and they tend to assert their freedom of action about all matters of Gulf security. For instance, the decision of Kuwayt to establish diplomatic relations with the Soviet Union in 1981 was not the expression of a desire to be a friend of an ally of the Soviet Union, but evidence that it wanted to pursue its own independent course in dealing with the superpowers. This policy of balance and moderation prompted Kuwayt to invite both the United States and the Soviet Union to escort its tankers in Gulf waters when they were exposed to threats by Iran in 1987. 'Uman and the United Arab Amirates, although unmistakably pro-Western, followed the example of Kuwayt in the establishment of diplomatic relations with the Soviet Union in 1985 and 1986. Other Gulf countries are likely to take similar steps in order to demonstrate solidarity in the search for peace and security in the Gulf.

The most significant attempt to deal with the problem of Gulf security by Gulf states was the establishment of the Gulf Cooperation Council (GCC) in February 1981. It is not our purpose to discuss the events leading up to the establishment of the GCC or its structure and processes, as these will take us far afield. Only the larger questions of peace and security, and the decisions taken to contain and, perhaps, ultimately bring the Gulf War to an end, will be dealt with here.[26]

The idea of a regional-security system is not new in the Middle East. It was implied in the pact of the Arab League and the Charter of the United Nations. The failure of the Arab League to provide adequate security seems to have prompted the Gulf countries to envisage a security system even before Britain decided to withdraw from the Gulf. The Arab Gulf countries were encouraged by their Arab neighbors to form an Arab union that would, in cooperation with Saudi Arabia, stand as a pillar in the Gulf to deal with security problems. Negotiations for a federal union among seven Arab countries—Bahrayn, Qatar, and the five Trucial Coast Principalities—led to the establishment of the United Arab Amirates in 1971, consisting of only the five Trucial Coast Principalities.[27] Not all the Arab Gulf countries, in the absence of a serious foreign threat, were yet ready to unite or cooperate for collective security on the regional level. Only after the Iraq–Iran War seemed to present serious threats to their very existence did they overcome local differences and dynastic rivalries to form an Arab Gulf security organization. Earlier, in 1976, when the foreign ministers of the eight Gulf countries—Iran, Iraq, Saudi Arabia, Kuwayt, Bahrayn, Qatar, the United Arab Amirates, and 'Uman—met in Masqat at 'Uman's initiative to discuss Gulf security and defense against foreign intervention, they were unable to agree on a common place. Nor did the Arab Gulf countries agree on a broader Arab regional cooperation when they met in Kuwayt later that year. Only five years later, after the Iraq-Iran War commenced, did six Arab Gulf states (without Iraq, although Kuwayt seems to have proposed its inclusion) meet at Kuwayt's initiative to form a Gulf union to deal with security within the framework of an organization whose function was to stress economic, social, and cultural cooperation. After preliminary meetings of the foreign ministers of the six Arab Gulf countries in February 1981, first in Riyad, and later in Kuwayt, the GCC Charter was finally drawn up. Three months later (May 26, 1981), the six heads of state met in Abu Dhabi to sign the charter, and the GCC formally came into existence.

The functions of the GCC, according to its charter, are embodied in the concepts of "cooperation," "coordination," and "integration" in economic, social, and cultural activities (the Preamble and Article 4) among members, leading to progress and development and, ultimately, to unity as envisaged in the pact of the Arab League. Thus in its structure and purposes, the GCC may be regarded as a "regional arrangement" in accordance with the United Nations Charter (Articles 52–54), designed specifically for the Gulf, although it was not intended to supersede but to supplement the functions of the Arab League.[28]

Since neither the Arab League nor the United Nations, despite many a resolution calling for a cease-fire and peaceful settlement, could do anything initially to stop the Gulf War, it devolved on either the major powers or the GCC to contain the war and deal with Gulf security problems. The major powers, including the United States and the Soviet Union, had already declared their strict neutrality toward the Gulf War. Under the Carter Doctrine, the United States pledged to defend the Gulf region only in the event of foreign (Soviet) intervention, and even the offer of military "cooperation" would take place by invitation of one or more of the Gulf states. Under these circumstances, the GCC was bound to deal not only with the defence requirements of its members, but also with Gulf security problems in general.

The GCC Charter refers to Gulf security only in broad terms, such as cooperation and coordination in "all fields" (Preamble), and "disputes" to be dealt with by a board called the "Commission for Settlement of Disputes" (Article 10). This commission is directly connected with the Supreme Council, composed of the heads of state, the highest authority in the GCC, to which all "recommendations or opinions . . . for appropriate action" are to be submitted (Article 10, paragraph 4). The Ministerial Council, composed of the foreign ministers of the six Gulf members, is the policy-formulating organ: it considers proposals, policies, and plans; receives recommendations and suggestions proposed by GCC members; and submits them to the Supreme Council for approval. The Secretariat of the GCC is the operating organ, whose headquarters is in Riyad, entrusted with all the administrative, integrating, and technical functions of the organization.

Although security matters are not spelled out in detail in the charter, military cooperation and coordination were often discussed from the very beginning, and the GCC began to pay greater attention to these subjects as the war dragged on. Both bilateral- and collective-defence plans were laid down and approved in principle, but their implementation was slow, or they were allowed to languish until the need for them arose. Bilateral-security agreements between Saudi Arabia and other GCC states—except Kuwayt—were entered into early in 1982 (reaffirmed and extended later), providing for the exchange of military information, training, and equipment; the extradition of criminals; and joint military exercises. The last went far beyond bilateral arrangements carried out under the auspices of a military committee set up to coordinate military operational efforts. In 1983, the GCC laid down a plan to create the Joint Strike Force, later called the "Peninsula Shield," when it held joint exercises in western Abu Dhabi later that year, intended to demonstrate the capability of GCC joint military efforts. These exercises were tried on a larger scale at Hafr al-Batin, in northeastern Saudi Arabia, where some 10,000 men from the 6 Gulf states participated in various forms of military demonstrations. The Joint Strike Force, modeled after the American Rapid Deployment Force, whose command was entrusted to a Saudi general, was at the outset experimental. However, the stationing of small units at Hafr al-Batin (near Kuwayt's border) may well prove its possible usefulness as a deterrent force in view of the concern over the expansion of the area of military operations in southern Iraq and the support it had received in official and unofficial circles.[29]

On all matters concerning domestic security—exchange of information about subversive activities, foreign travel, passports, and the like—there seem to have been no essential differences of opinion. Subversive activities reported by one country to another will always be helpful to avoid or abort sudden disturbances. A case in point is the exchange of information noted earlier between Saudi Arabia and Bahrayn concerning those involved in the attempted coup in Bahrayn in December 1981.

On matters concerning regional security, there had been differing views even before the GCC came into existence. For instance, Kuwayt, surrounded by countries much larger than itself, stressed diplomacy as a means to settle disputes, while ʿUman, on the opposite side of the Gulf, preferred to depend primarily on military power. Saudi Arabia, the major Arab power in the Gulf, advocated a combination of the two methods. After the establishment of the GCC, the Saudi method of "balance and moderation" seems to have become acceptable to all GCC countries in principle, although not without qualifications, each in accordance with its requirements and resources.

But security cooperation was not confined to strategy and joint military efforts. It also included negotiations with foreign powers to obtain military assistance and weapons. There were three major sources of arms for GCC members: the United States, Western Europe, and the Soviet Union. To coordinate their needs and military plans in accordance with their weapons systems, it was deemed necessary to obtain arms from one primary source. Dependence on one primary source, however, with all its military advantages, may become a handicap if political differences develop between the supplier and the purchaser, leading to the unexpected interruption of arms sales. Diversification, it was tacitly agreed, would be necessary in order to avoid dependence on one source of supply. Saudi Arabia, facing congressional opposition because of legislators' concerned with the security of Israel, decided to diversify its arms by purchasing combat planes from Britain and France, although it preferred to purchase all its arms and equipment from the United States. Apart from Saudi Arabia, the United States provided limited supplies of arms to Bahrayn and the United Arab Amirates. ʿUman, which traditionally has enjoyed a close friendship with Britain, preferred to purchase British arms and employ British military personnel to train its army. Apart from Iraq, Kuwayt is the only Arab Gulf country that has purchased arms from the Soviet Union.

Perhaps another area of military cooperation would be the establishment of a joint arms industry. Egypt and Iraq, although their modest beginnings proved in the main experimental, are today the two countries that have the greatest potential to develop national arms industries. The GCC seems to have seriously considered the need for a national arms industry, with the possible cooperation of Arab and Islamic countries like Egypt, Turkey, and Pakistan. Although funds have been earmarked, it will take a long time before an arms industry would be ready to provide highly developed weaponry to meet Arab military needs.

With regard to the Iraq-Iran War, the major Gulf security problem, the GCC sought to deal with it on two levels: first, to contain the war by defensive measures, and, second, to end the war by bringing the two sides to the negotiating table. The first, as noted earlier, was dealt with either unilaterally or collectively. Before 1982, when Iraqi forces were still fighting on Iranian territory, the war seemed to be a conflict between two littoral Gulf states, although air raids often "strayed" to hit civilian centers or interrupt shipping in the Gulf. Strictly speaking, the position of the GCC may be described as neutral, as it has taken a defensive attitude and refrained from providing arms to either side, although its sympathy was obviously on Iraq's side on ideological grounds. Some members of the GCC, especially Saudi Arabia and Kuwayt, have unilaterally extended financial and even some logistical assistance; but each did so on its own and not on behalf of the GCC.

The other objective of the GCC—to bring the war to an end—is the prerogative of any third party, provided it seeks to achieve it by diplomacy and not by resort to force. True, the GCC laid down plans for acquiring arms and conducting military exercises; but these were defensive measures and were by no means intended to assist one side or the other. Diplomacy was and still is the only instrument that the GCC advocates, either alone or in cooperation with other powers or councils, to persuade the two sides to accept a cease-fire and settle their differences. Indeed, critics have reproached the GCC that even its diplomatic endeavors were benign and did not use the full resources it has at its disposal. Nevertheless, the attempts that had been begun as early as 1982, have not yet been given up. In June 1982, when Iraq announced that it had decided to withdraw its forces from Iranian territory, Iran refused to accept the Security Council resolution calling again for a cease-fire, and carried the war into Iraqi territory a month later. The Arab countries, at a summit meeting in Fas (Fez) in September, branded Iran as the aggressor; but the GCC took no action, save that some of its members expressed their concern. When, however, the Iranian offensive failed and Iraq demonstrated its ability to repulse Iranian attacks, the GCC began to realize that the war might be prolonged. The only way to bring it to an end, it was argued, was by diplomacy. Some seem to have suggested that an agreement reminiscent of the Marshall Plan, a reparation for both sides, might be tempting to remind Iran and Iraq that the time had come to divert their attention from destruction to construction.

The idea that some form of reparations might play a role in resolving the conflict was broached early in May 1982, by the Iranian ambassador to the United Nations. Following the shift in the military confrontation between Iraq and Iran in June and July 1982, several meetings of the Ministerial Council of the GCC were held, and conciliatory statements to Iran were reported in the press. It was informally reported that the GCC was prepared to offer Iran an amount between $10 and $25 billion for reconstruction, if a cease-fire were to take place along the front. Although the offer of funds was publicly denied, the request for a cease-fire was not. Nor did Iran reject the GCC proposal out of hand; indeed, men in high offices, including President Khami' ini, made statements in which it was indicated that Iran's losses in the war amounted to over $150 billion. It was also remarked that reparations should be paid by Iraq and not by other parties. But the GCC seems to have made no further offers, although attempts at mediation were not discontinued.[30]

During the following two years, from 1983 to 1985, the foreign ministers of Kuwayt, Saudi Arabia, and the United Arab Amirates undertook mediatory missions to restrain the ferocity of the war—the war of attrition on land and sea, including the bombing of refineries and oil terminals, such as Kharj Island, and attacks on tankers[31]—and they seem to have made some progress following visits to Tihran, Baghdad, and other Arab capitals, as transpired in some of the official statements, such as the communiqué of the Ministerial Council meeting at Masqat (Muscat), ʿUman, on November 6, 1985.[32] But no change in the official position of Iran was ever made. On the contrary, Iran often warned GCC countries to distance themselves from and stop their financial support of Iraq. Moreover, there were several terrorist incidents in GCC countries, especially in Kuwayt and Saudi Arabia, resulting in personal injuries. On May 25, 1985, the Amir of Kuwayt, Shaykh Jabir al-Ahmad al-Sabah, narrowly escaped from a suicide bombing attempt on his life for which the Islamic Jihad claimed credit. These subversive activities, of course, have continued.

The failure to bring the two neighbors to the negotiating table prompted the GCC members to appeal to the major powers, either individually or collectively through the United Nations, to exert pressure on Iran to accept a cease-fire. By its action, the GCC demonstrated that in a crisis, regional endeavors are inadequate to maintain peace and order. But before we deal with the question of the involvement of the major powers in the Gulf War, we should discuss how they were drawn into it.

No study of the Gulf War would be complete without a few words about the role of oil in war. Indeed, the impact of oil on war can hardly be over-emphasized. Apart from being an essential source of energy for modern warfare, it is an important commodity in great demand in all the industrial countries of the world. Any attempt at interrupting the flow of oil to Western markets is likely to invite foreign intervention. Since both Iraq and Iran are oil-exporting countries, both have sought to use oil as a weapon in their military confrontation. This situation has tended to prolong the war, and the two countries are not unaware that the outcome of the war will depend not only on their military strategy and potential, but also on oil as a weapon in diplomacy. In the equation between the two—diplomacy and war—the latter cannot possibly be won without the aid of the former.

It has been maintained by some geopolitical experts that in the long run, the outcome of the Gulf War is likely to be in favor of Iran because of its superiority in human resources and geostrategic depth, presumably on the assumption that the Iranian leaders would not hesitate to turn to diplomacy in their relationships with the major powers. In the experiences of mankind, diplomacy has always been employed to achieve *raison d'état* (national interests, values, and ideology), but today the geopolitical experts seem to have reduced *raison d'état* to the variables of geography and foreign policy, without regard to other concerns, religious or otherwise, that are still important to leaders who have no high regard for secular and positive standards. In order to understand the attitude of the revolutionary leaders in Iran toward the world powers, these subjective variables should be taken into serious consideration.

When the war started, Iraq's well-wishers held that the war would not last long because of Iraq's military superiority and the disarray in the Iranian army following the revolution. Had Iraq pursued its initial victories more vigorously and the disaffected Iranian provinces responded by uprisings, Iraq might have had a good chance of winning a quick victory. But Iran was not slow to attack Iraq in other fronts. No sooner had the Iraqi forces crossed the frontier than Iran destroyed Iraq's oil terminals at the head of the Gulf— Faw, Port Bakr, and Umm Qasr—and stopped the flow of oil through Gulf waters. The Syrian trans-desert pipeline, although remaining in operation at the beginning of the war, was shortly afterward closed. The only pipeline still open to Iraq was through Turkey. Iraqi oil exports thus at once dropped from some 3.5 million barrels a day before the war to only 650,000 barrels a day after the war had begun. Before it was able to make alternative arrangements to increase its oil exports, Iraq had to import oil for its military requirements and to fall back on its savings in hard currency in foreign banks to meet budget deficits. Moreover, Iraq was also supported by Arab Gulf countries, especially Saudi Arabia and Kuwayt, which contributed over $1 billion a month before Iraq was able to increase its oil exports two years later. In June 1982, Iraq withdrew its forces from Iran, hoping that a ceasefire and settlement of the differences between the two countries by peaceful means would be acceptable to Iran. But the Iranian regime, contrary to expectations, began to take the offensive. The war passed from its first stage, in which it was confined to land operations, to its second, in which it was extended to a wider area in which first the Gulf countries and later the major powers were gradually involved. During the latter stage, oil was used more effectively as an instrument of diplomacy.

Before it could play its oil card, Iraq had to make alternative arrangements for the oil terminals it had lost at the beginning of the war. It achieved that objective in two stages: first, by constructing new pipelines and expanding the capacity of already existing ones in friendly neighboring countries; second, by attempting to raise the quota of oil production allocated to it by

OPEC. The objectives were closely interrelated, as the implementation of the latter was dependent on the capacity of pipelines to deliver oil at the terminals.

With regard to the first objective, Iraq began to expand its oil-export capacity through Turkey and Syria. Syria was persuaded to allow the export of oil by an agreement in March 1982, but the arrangement did not last long, as the pipeline was closed three months later. Turkey proved more reliable, and the capacity of the pipeline from Kirkuk to the Turkish port of Ceyhan, on the Mediterranean, was increased from 650,000 barrels a day to almost 1.3 million barrels a day by 1983. Another agreement, concluded in 1985, provided for the construction of a pipeline running parallel to the first, to be completed in 1987, which would raise the oil-export capacity to another million barrels a day. A second project, agreed to with Saudi Arabia in 1984, was to tie in the existing Saudi east–west pipeline with the Iraqi southern oil fields at Zubayr, which began to flow in 1985 at an initial rate of 300,000 barrels a day and increased to 500,000 barrels a day in 1985. The construction of a parallel east–west pipeline was agreed on, to be completed by the end of 1987, which would allow Iraq to export 1.6 million barrels a day through Saudi Arabia. The total Saudi and Turkish pipelines, when they are completed, are expected to increase the oil export capacity to almost 4.5 million barrels a day. Finally, a third project is a 540-mile pipeline through Jordan, running from Iraq's northern oil fields to the Red Sea port of 'Aqaba, with a capacity of 1 million barrels a day. This project had already been agreed on between Iraq and Jordan (Iraq would pay the cost of construction of the pipeline running on its territory, roughly about 200 miles, and Jordan would pay its share for construction of the rest of the pipeline, roughly 350 miles), but negotiations for the construction of the pipeline by American companies (under the supervision of Bechtel Corporation) have long been delayed by the stipulation that the repayment obligations would lapse in the event that Israel interfered with the construction or operation of the pipeline. Should Syria decide to permit the flow of oil from Iraq, another 500,000 barrels would be added to Iraq's oil-export capacity. If all the pipelines—the Turkish, Syrian, Jordanian, and Saudi lines (not counting the Iraqi terminals in the Basra Province)—were to be used, the Iraqi export capacity would exceed Iraq's own requirements and would be more than it had ever hoped to export before the war.

The second objective is to increase Iraq's export quota of oil allocated by OPEC. At the beginning of the war, Iraq's capacity to export oil could hardly exceed 650,000 barrels a day, although it was allowed a quota of 1.2 million barrels a day in principle. Iraq's quota remained unchanged until 1985, when it applied for an increase, but no decision was made. In 1986,

Iraq again applied for an increase, as Iran's quota had been raised to 2.5 million barrels a day. President Saddam Husayn, in a letter to Iraq's representative to OPEC, demanded equality with Iran, arguing that Iraq's requirements should be measured not only by demographic standards, but also by its defence requirements. Since no decision has yet been taken to meet Iraq's request, Iraq informed OPEC that it will not be bound by OPEC allocations. At any rate, the Iraqi representative stated at OPEC's last meeting, in 1987, that Iran's oil exports have not always been consistent with OPEC's allocations. Iraq seems to be determined that its oil quota should be equal to Iran's in principle.[33]

While Iraq was still grappling with the problem of increasing its oil-export capacity, it lost no time in trying to make clear its intention to bring the war to an end by peaceful means in accordance with United Nations resolutions. It also sought to enlist the support of the major powers—indeed, all the powers that have an interest in the Gulf region—in bringing pressure to bear on Iran to accept a cease-fire. However, before the United States and the Soviet Union began to tilt toward Iraq, they had their own dealings and frustrations with Iran (although the Soviet Union has not given up all dealings with Iran). Iraq at once seized the opportunity of American and Soviet disenchantment with Iran, and its oil diplomacy began to work in its favor. Tariq ʿAziz, its new foreign minister, known for his flexibility and resourcefulness, visited Moscow in July 1982, shortly after he took up his post, and succeeded in obtaining a Soviet agreement to resume arms shipments to Iraq. The Soviet Union agreed to lend Iraq over $2 billion in 1983 and another $2 billion in 1985, to be repaid in crude oil. ʿAziz also visited France, and his negotiations resulted in agreements, concluded in 1983 and 1985, to extend credit for the purchase of aircraft (Mirage and Super Etendart) and new missiles (Exorcet), which Iraq had begun to purchase from France earlier. In November 1984, Iraq resumed diplomatic relations with the United States. This step, allowing Iraq to purchase American technology not prohibited to combatants, was the signal of American readiness to cooperate with Iraq. Since American policy has always been in favor of the United Nations Security Council's calls for a cease-fire and the settlement of conflicts by peaceful means, the resumption of diplomatic relations has been described in the American press as a "tilt" toward Iraq.

But the tilt seemed at first merely a symbolic gesture for the resumption of diplomatic relations between the two countries without necessarily altering the strict neutrality of the United States in the war. Despite efforts to enhance American-Iraqi cooperation, pressure groups friendly to Iran and thus opposed to the tilt argued that Iran was geopolitically more important than Iraq, which made policy makers remain hesitant about altering American strict neutrality. Meanwhile, public concern about American hostages in Lebanon was more frequently expressed in the press. Terrorist activities in Lebanon and other countries were not new events in the Middle East. But they were discussed in relevance to Iran's influence over Shiʻi terrorists in Lebanon, although Iran had been inciting Shiʻi subversive activities in Iraq and other Gulf countries since the Islamic Revolution began in 1979. It was suggested to President Reagan that a friendly gesture to Iran—the delivery of weapons and spare parts—might induce "moderates" to resume Iran's relationship with the United States. The contact with Iran, however, carried out through covert channels, failed to achieve the purpose of pro-Iranian policy makers, since the Iranian "moderates" who accepted the delivery of arms—their sole purpose in the deal—ordered the release of only three hostages, and their attitude toward the United States did not seem to differ from that of the "extremists." The ill-advised Iran venture, ironically called Irangate, is still incomplete and the full story has yet to be revealed.

It was now an opportune moment for Iraq, to persuade the Reagan administration to make the "tilt" toward Iraq more meaningful. Two events may be said to have prompted the United States and the Soviet Union to take more direct action in the Gulf War. Early in 1987, Kuwayt, whose oil tankers had become the target of increasing attacks, invited both the United States and the Soviet Union (and probably other powers) to escort its tankers through the Gulf. Perhaps it was not surprising when the Soviet Union, known for its desire to have a presence in the Gulf, accepted the invitation even before the United States had yet considered the matter. Faced with this challenge, the United States could not possibly turn down the invitation from a Gulf country known for its moderation and balance. While the Reagan administration agreed in principle to reflag eleven Kuwayti tankers and to escort them through the Gulf, it was warned by Congress not to get more deeply involved in Gulf affairs. Reagan, determined to ensure the free flow of oil, appealed to European allies to participate so that the undertaking would be international and not reduced to a superpower rivalry. Britain and France, whose warships had already been in the Gulf to protect their tankers, endorsed Reagan's action in a unilateral, not a collective capacity. A collective action carried out by the world powers through the United Nations would surely have a greater political and moral impact on all the Gulf countries and bring about greater cooperation to end the war. But the mandatory resolution of the Security Council, made on July 20, 1987, to which we shall return, has yet to be implemented in the same spirit of cooperation among the major powers that brought it into being.

The second event was the attack on the U.S.S. *Stark*, inadvertently carried out by an Iraqi pilot on May 17, 1987. It alerted the American Naval

Command that the mere presence of an American warship in the Gulf would not be adequate to protect tankers without American naval and air protection in the event of an engagement with any attacking force. The presence of an American military force in the Gulf, raising the prospect of possible closure of the Strait of Hormuz and other retaliatory measures, such as the mining of Gulf waters and attacks on tankers by missiles, have virtually brought American forces face to face with Iran. The official American policy is to pursue so-called Operation Staunch—ensuring free navigation in the Gulf by escorting the tankers of Kuwayt. If challenged, however, will the American air and naval forces react by attacking Iranian installations?

Intervention in such a situation would become almost inevitable as a measure of defence. With regard to bringing the war to an end, American policy aims at seeking cooperation with the other major powers through the United Nations. It is, indeed, the safest way not only to reduce the superpowers' rivalry, but also to enlist the cooperation of Gulf countries in the endeavors to end the war.

# XII

# Conclusion

Before concluding with a discussion of the prospects for the reestablishment of peace and order in the Gulf, perhaps a summary of the cumulative drives that brought about war and chaos in that region might be useful. For the military confrontation between the two littoral neighbors at the head of the Gulf, although stalemated, might come to an end at any moment; but unless the issues that ultimately led to war are resolved, resort to violence will be exceedingly difficult to avoid, regardless of the regimes in Iraq and Iran.

The Gulf War began as a conflict between Iraq and Iran over border claims and violations before other Gulf countries were directly or indirectly drawn into it. The root cause of the conflict, however, was and remains the Sunni–Shi'i confessional controversy, which divided the house of Islam into two major religious communities. Before the rise of Persia as a Shi'i state at the beginning of the sixteenth century, the schism in Islam was communal, or horizontal, not accompanied by territorial segregation; but after the secession of Persia from Islamic unity and its adoption of Shi'ism as the official state religion, the schism became territorial, or vertical, and set in motion a decentralization that had been in progress for centuries and that led to the fragmentation and eventual breakup of the house of Islam into several independent states.

Persia's adoption of Shi'ism led to a conflict with the Ottoman Sultan not only because the Sultan had become the spokesman of Sunni followers, but also because Persia's action betrayed a latent ethnocultural identity in the era preceding nationalism that aggravated the confessional controversy. In its conflict with the Ottoman Empire, Persia sought to achieve two main objectives: first, to gain an independent status, separate from the rest of the Sunni world; second, to extend its domination over the northern Gulf region, including the southern areas of Iraq, where the Shi'i holy sanctuaries exist and whose inhabitants were and still are predominantly Shi'i followers.

Persia's ambition ran contrary to Ottoman policy, supported by Sunni Arabs, which aimed at the reestablishment of Islamic unity. The ensuing

159

Perso-Ottoman rivalry, leading to continuing confessional tension and military confrontations, exhausted the power and resources of the two warring empires and reduced their prestige and influence in the Muslim world. Their decline, coinciding with the rise to power of rival European neighbors, induced them eventually to come to terms with life and recognize each other's existence as sovereign Muslim states, relegating religious differences to the domestic level. The new map of the Muslim world, revealing the realities of political conditions, was finally given official expression under the Treaty of Ardurum (Arzurum) in 1847. The next step was to draw up political frontiers and establish permanent landmarks as evidence that each side was prepared to respect the sovereignty and territorial integrity of the other. After long and tedious endeavors, the demarcation (except the final stage on the Shatt al-'Arab, from the city of Basra to the Gulf) was almost completed when the First World War broke out.

Following the First World War, when the Ottoman Empire collapsed, Iraq as a successor state inherited the legacy of regional fragmentation, confessionalism, and all other unresolved issues that had arisen between the Persian and the Ottoman Empires. But as new modern states, both Persia (renamed Iran in 1935) and Iraq joined the League of Nations and sought to conduct their neighborly relations and settle their differences by peaceful means in accordance with the League Covenant and the rules and practices of International Law and diplomacy. The unresolved border problems inherited from Ottoman times were finally settled in 1937, and both Iraq and Iran became signatories to the Four-Power Security Agreement (called Saʿdabad Pact) promising to maintain peace and order in the Middle East. By the end of the inter-war years, Iraq and Iran were able to iron out major differences, and their cooperation in the pursuit of peace and stability in the region seemed to be an ongoing historical process that could not be interrupted or reversed.

The Second World War, however, brought in its train new social forces and demands that placed Iraq and Iran on opposite sides on almost all matters concerning security and bilateral relationships. The ensuing conflict was not confined to Iraq and Iran; indeed, it extended to several other Middle Eastern countries. Today, in the escalating Gulf War, the whole region is involved in conflicting demands and claims.

What were the causes and drives that led to this situation?

Apart from the ideological rivalry between the two superpowers, into which many countries in the Third World were drawn, Iran and Iraq began to pursue entirely different goals when their income from oil became available for development. From the early postwar years, the small ruling classes in both countries, jealous of the power they had at their disposal, were reluctant to share it with the public, although the parliamentary system in both countries, providing for public participation, was prescribed in their written constitutions. Considering authoritarian rule to be the most effective form of governance, the ruling classes relied, in the last analysis, not on public support through parliamentary elections, but on the army and the police as the shield of their regimes. In Iran, the military by tradition developed deep loyalty to the monarchy and continued to protect it, even against popular protests, until it was overthrown by the Islamic Revolution in 1979. But in Iraq, the military was influenced by entirely different social forces and often came into conflict with the monarchy. After several attempts, the military finally seized power and abolished the monarchy in 1958. Not only did military rule in Iraq pursue development based on radical social and economic doctrines, but it also advocated national objectives perceived by the Shah of Iran and the Western powers to undermine regional and global security. As a result, for almost two decades, from 1958 to 1975, Iran was in constant conflict with Iraq, a conflict manifested either in its demand to alter the border line in the Shatt al-'Arab or in its support of opponents and dissident groups in their struggle for power against the regime.

In 1975, when the Shah of Iran and Saddam Husayn, Vice Chairman of the Revolutionary Command Council of Iraq, finally met at Algiers, they came to a quick understanding. Each seems to have made it crystal clear that his country's regional aims were in agreement with those of the other to maintain peace, order, and good-neighborly relations. No less important, Saddam Husayn pointed out, the Ba'th regime in Iraq, not unlike the Shah's in Iran, sought to achieve national objectives (that is, independence, territorial integrity, and security). To meet the Shah halfway, Saddam Husayn proposed that Iraq would concede Iran's claim to the thalweg as the water boundary between the two countries, provided that Iran returned the three central land sectors to Iraq. Finally, both the Shah and Husayn agreed to suppress subversive activities originating in one country against the other and put an end to interferences in domestic affairs. The Algiers reconciliation, not unlike the 1937 arrangement, not only spared Iraq from threats and subversive activities, but also gave the Shah a free hand to deal with his own domestic and security problems. The Shah's rule, however, had become so high-handed that groups on both right and left felt compelled to enter into an uneasy alliance to rid the country of his regime. To his surprise, the army, the traditional shield of the monarchy, failed to come to his rescue. The Shah's throne, with no support even from allies, fell by its own weight. As a result, the Algiers reconciliation, not unlike the 1937 arrangement, proved to be a twilight, as the Iranian Revolution showed little or no respect for international obligations.

Iraq, still awaiting the return of the three central land sectors, offered to

recognize the new Iranian regime, provided that the provisions of the Algiers Agreement would be respected. But the leaders of the Iranian Revolution were thinking about other matters. In both public and private statements, they made it clear that they were not interested in such matters as
the dispute about frontiers, but in the larger questions of establishing Islamic
institutions and asserting Islamic standards. Since Muslim rulers in neighboring lands, the Iranian leaders declared, had fallen under foreign and corrupt influences, it was the duty of believers to warn them—even to call for
their downfall—in order to reestablish the rule of Islamic law and justice.
Just as the jihad had often been invoked in the past against rulers who violated the sacred law, the Ayat-Allah Khumayni called for "exporting the
Revolution" as an instrument of the jihad to rid believers of oppressive and
corrupt rulers.

The Iranian leaders have thus undertaken the task of extending the bene-
fits of the Iranian Revolution to other Islamic lands whenever they feel that
the regimes across their borders seem to be straying from Islamic standards.
The initial response outside Iran was not only in Shiʿi communities, as Khumayni's teachings on the whole did not stress denominational differences,
especially in his writings preceding the Iranian Revolution. His followers in
Iran, however, have consciously asserted Shiʿi doctrines and traditions that
often betray Persian ethnocultural bias and have alienated conservative Sunni
followers. The new Iranian regime seems to follow the precedent established
by Shah Ismaʿil al-Safawi in the sixteenth century, advocating Shiʿi teachings and traditions rather than the principles and concepts of government
laid down by the Prophet and implemented by his early successors. It has
thus renounced the doctrine of the separation of religion from the conduct
of state in foreign affairs, to which former Iranian regimes and other Muslim
countries subscribed as members of the modern communities of nations. By
invoking the Imam's authority, Iran's leadership has claimed the allegiance
of believers not only in Iran, but also in other lands. More specifically, by
asserting the Shiʿi Imamate as the supreme authority, it implicitly has called
on all Shiʿa to play the role of a fifth column by undermining and ultimately
overthrowing regimes to which they are expected to owe allegiance. Iran
has sought to achieve its goals not only by peaceful methods—including the
media, Shiʿi institutions outside Iran, and the annual pilgrimage—but also
by subversive activities and military incursions into the territories of its
neighbors. Small wonder that today many Muslim scholars outside Iran,
including Shiʿi scholars of Arab descent in Iraq, have rejected the claim of
Shiʿi scholars in Iran to speak on behalf of believers about what Islamic
standards ought to be.

By contrast, the structure of politics in Iraq has been subjected to varied streams of thought and has followed a different line of political development ever since the country came into existence following the First World War. After a short British tutelage, under which it adopted Western concepts of law and authority, Iraq emerged as an essentially modern secular state. True, its political system passed under military rule after independence and only recently did the military begin to recede, but the general character of the political system remained secular. No less important, the Iraqi leaders have adopted social and economic doctrines stressing modern secular standards, derived partly from Iraq's own experiences and partly from the experiences of other societies. Since the Islamic institutions in Iraq were and still are divided, some under Sunni and some under Shi'i clerics, religious resistance to secular trends in Iraq were less influential than in Iran. Moreover, all educational institutions are under state control, the Sunni clerics, like civil servants, receive their salaries from the state. But most Shi'i educational institutions prefer to receive their endowments from private sources, and Shi'i clerics who receive their pay from private sources exercise a greater influence on their followers than do Sunni clerics. Nonetheless, because of the subordination of most religious organizations to the state, the secular character of the political system remains immune to clerical influence.

Owing to the secular character of the Iraqi regime, often branded in the Iranian media as infidel and antireligious, the Iranian leaders sought to undermine it in the eyes of its people and called for its downfall. Since the Iraqi leaders felt secure internally, their main concern was with the frequent Iranian incursions and violations of Iraq's frontiers by Iranian forces. Above all, the Iranian regime had not yet given up the three central land sectors that the Shah had pledged to return. Before it took retaliatory action, the Iraqi government addressed several notes to Iran calling for negotiations to settle pending frontier issues in accordance with the Algiers Agreement and the treaty of 1975, which had been duly signed and ratified. Since no replies were received and violent violations of the frontier were not stopped, Iraq denounced both the Algiers Agreement and the treaty of 1975 on the grounds that Iran had failed to abide by the terms of the treaties and had refused to negotiate the nations' differences in accordance with the procedure provided by these instruments. Since they stipulate that the violation of any provision by either party renders their terms null and void, Iraq felt that its action was justified. True, Iran protested Iraq's denunciation, presumably implying that it was still bound by the terms of the treaties, but refused to negotiate with the Iraqi regime.

Unable to persuade the Iranian leaders to settle their differences with Iraq by peaceful methods, Iraq should have referred its dispute with Iran to the Security Council (as it had referred its dispute with Iran to the League of

Nations in 1934), and Article 51 of the United Nations Charter should have been invoked, before taking any unilateral action. Since Iran had already resorted to violence—it began to bomb towns and villages within Iraqi territory as early as September 4, 1980—Iraq felt compelled to defend itself. Iraq sought to justify its action before the Security Council on the grounds that it had had to take "preemptive" measures in defence of its territorial integrity without invoking Article 51. Preemptive action is a recent doctrine advocated by some jurists, but it has not been accepted as a principle of law. For this reason, when the Iraqi forces crossed the frontier on September 21 and 22, 1980, Iran charged Iraq with aggression, although Iran itself invaded Iraqi territory later (after Iraq had withdrawn its forces in 1982), claiming that it pursued the Iraqi forces in self-defense. Iran has also refused to abide by Security Council resolutions calling for a cease-fire and settlement of differences by peaceful means. Iraq has accepted not only all Security Council resolutions calling for a cease-fire, but also all offers of good offices by third parties, which Iran has rejected out of hand.

Three kinds of mediation and good offices have been attempted. First, international and regional organizations, such as the United Nations and the Arab League, have become involved. Some, like the United Nations Security Council, have passed over a dozen resolutions calling on both sides to settle their differences by peaceful means. On July 20, 1987, the Security Council unanimously passed a resolution ordering Iran and Iraq to accept a cease-fire. While Iraq accepted the resolution at once, Iran raised several objections and demanded certain conditions before it would accept it. The Security Council, however, has yet to apply punitive measures, economic or otherwise, to enforce its mandatory resolution on the recalcitrant party refusing to accept the cease-fire.

Second, ad hoc or permanent commissions or delegations were established to inquire into the practical and most appropriate methods to bring the war to an end. The Palme mission, appointed by Kurt Waldheim, Secretary-General of the United Nations, in November 1980, and the Good Offices Commission, established by the Islamic Conference Organization in January 1981, are but two notable examples. Although most of these missions have ceased to exist, the United Nations and the Islamic Conference missions are still active and may well eventually succeed in achieving the suspension of hostilities.

Third, men in high offices (some in private capacities) sought to impress the need on heads of state and government in friendly relations with Iranian leaders to end the war. Although visits by persons in their private capacities have not been publicized, these endeavors to influence opinion might well be more far-reaching than those of others in official capacities. But the attempts of persons in official or private capacities have not yet succeeded, as they must be coordinated with one another by high authorities if they are to be effective. In practice, however, the peace offensive of one mission is not infrequently frustrated by the attempts of another to seek accommodation with the Iranian regime.

Why have all these attempts to end the war failed? Some of the regional organizations that might be capable of dealing with the Gulf War suffer from certain self-imposed limitations. A case in point is the Arab League. By its very nature, it is an Arab and not, strictly speaking, a comprehensive "regional arrangement." While it could deal with disputes among Arab Gulf members, such as Iraq's claim to the sovereignty of Kuwayt in 1961, it has proved unable to resolve issues outside the scope of its pact, since one of the principal Gulf states—Iran—is not and cannot become a member. Nor did the Islamic Conference Organization prove to be a suitable channel, even though Iran is a natural member, as this organization is composed of countries extending from the Atlantic to the Pacific oceans and representing more than one ethnocultural region with conflicting interests. As an Islamic organization, it can in theory invite both Iran and Iraq to present their conflicting views. But when the Islamic Conference met in Ta'if (1981) and Kuwayt (1987) to discuss, among other things, the Iraq-Iran War, Iran refused to attend both conferences—the first for inviting Iraq while its forces were committing aggression on Iranian territory, and the second for meeting in Kuwayt, which Iran considers to be not neutral in the war.

The GCC, designed specifically for Gulf security, is perhaps potentially the most suitable regional organization for dealing with the Gulf War. However, owing to its present structural and procedural limitations, it is not expected to cope with all regional problems. Nevertheless, it has already discussed some defensive measures, to be carried out individually or collectively. In almost all its meetings, it has been reported, the subject of the Gulf War is touched on, and some specific proposals for mediation were formulated and presented to both sides. But because Iraq and Iran are not members of the GCC, it is deprived of the opportunity to deal with the problems directly relating to the parties concerned. If the GCC is ever to become an effective organ to maintain peace and security, the door of admission must be thrown open to all Gulf countries.

As a Gulf security system, Iraq and Iran should in their own right as Gulf countries become members of the GCC. Before the outbreak of the war, it was almost taken for granted that those two states would be natural members of any regional council dealing with Gulf affairs. Indeed, Iraq had attended an Arab Gulf conference held at ʿUman in 1979. However, the security of the Gulf depends not only on membership of Gulf states for their geographical location, but also on geopolitical grounds of some countries, such as the two Yamans. Even some of the Red Sea countries, apart from Saudi Arabia, might be important were the Gulf exposed to danger from their side. Just as NATO, a security system originally designed for the Atlantic community, is now composed not only of Atlantic but also of Mediterranean countries (such as Italy, Greece, and Turkey), so the GCC might extend membership to countries outside the immediate Gulf community. In addition to the two Yamans, Jordan, which borders Saudi Arabia and Iraq and has an outlet to the Red Sea, would be strategically invaluable for the security of the Gulf, and the inclusion of all three may well enhance the stature and power of the GCC. For, if either of the two Yamans or Jordan were threatened from neighbors, the security of Saudi Arabia, Kuwait, and Iraq would be endangered.

As it stands today, the GCC is composed of countries whose regimes, outlooks, and policies are perhaps more harmonious than those of other Gulf countries, like Iraq and Iran, whose structures and socioeconomic systems are a mixture of modern and traditional patterns. The GCC states, unaffected by radical ideologies, prefer to pursue balanced and moderate policies in their relationships with neighbors as well as with rival powers outside the Gulf. For example, the Iranian leadership has repeatedly warned that if the Arab Gulf countries do not distance themselves from Iraq and stop giving Iraq economic assistance, they will be exposed to danger. But no GCC country, save Kuwait—accused by Iran to have been deeply involved in assisting Iraq—has yet been seriously hurt or its security jeopardized, notwithstanding that a number of attacks on tankers have taken place and free navigation through the Gulf has been threatened. Nor has the threat to close the Strait of Hormuz to shipping materialized, partly because of defensive measures taken by Gulf countries, such as the building up of an adequate air defence force, and partly because it prompted the major powers to provide escorts and protection against violations of free navigation on the high seas. But in the long run, responsibility for peace and security must rest with the Gulf countries themselves. To achieve this goal, the GCC must make possible the inclusion of all countries directly concerned.

Since the Gulf countries have not yet been able to end the war through collective regional endeavors and the flow of oil seems more seriously threatened, Iraq has called on the major powers to persuade Iran to accept a cease-fire. At the outset, the two super powers, each for its own *raison d'état,* tried unilaterally, but in vain, to bring about peace provided their own interests were protected. Iran's refusal to heed either one on ideological (Islamic) grounds prompted the United Nations to reactivate its efforts to bring the two warring parties to the negotiating table.

Two events, whether by accident or design, assisted Iraq's diplomatic endeavors. The first was Kuwayt's request that both the United States and the Soviet Union escort its oil tankers through the Gulf. Some have speculated that Iraq may have inspired Kuwayt to invite the Soviet Union to escort its tankers in order to draw the United States into Gulf affairs and, ultimately, exert pressure on Iran to end the war. But there is no evidence that the attack on the USS *Stark,* the second event, was a deliberate act. Iraq's inadvertent attack on the *Stark* suddenly alerted the Reagan administration to the possible Iranian threat to an inadequately protected American military force in the Gulf. It has been conjectured that the attack on the *Stark* may have been a deliberate act by Iraq to bring home to the American public the seriousness of the danger to Western interests in the Gulf. The Reagan administration, however, facing congressional opinion that Gulf oil is not a primary concern to the United States, has often tried in vain to make clear to the public and to congressional opponents the ultimate danger of Soviet intervention. These two events—Kuwayt's invitation to the Soviet Union to escort its tankers and Iraq's inadvertent attack on the *Stark*—may well have added weight to Iraq's argument in favor of the involvement of the major powers in the process of ending the war. Even before the United States formally agreed to escort Kuwayt's tankers, England and France—followed by Italy, Belgium, and Holland—had dispatched military ships to escort their commercial vessels, and other powers may join to reassert order and free navigation in the Gulf.

The settlement of conflicts in a region like the Gulf, rich in oil and important for its geopolitical and strategic location, should not be the responsibility of the major powers alone, but of all other nations concerned, as regional conflicts are likely to be the source of global conflicts. Because of the potential global dangers, such conflicts should be dealt with by the United Nations. Indeed, the Security Council has already passed over a dozen resolutions, including some by the General Assembly, calling on both sides to accept a cease-fire and settle differences by peaceful means. But those resolutions were not mandatory and were accepted by one side and rejected by the other. Since after seven years, all endeavors have been ineffective, mandatory acts are necessary, as intervention by rival powers might complicate and prolong the war. Indeed, some countries have already warned about the dangers of intervention by the superpowers. In these circumstances, the United Nations, acting on behalf of the community of nations, is perhaps the safest pathway to peace.

The proposal to revive action by the United Nations began at the regional level. At a meeting of the council of the Arab League early in February 1987, a seven-man delegation was appointed to the permanent members of the Security Council to impress on them the gravity of the developing crisis in the Gulf—increasing frequency of attacks on tankers, violation of free navigation, and attacks by missiles on oil installations in Kuwait—and its possible spread into more serious conflicts outside the Gulf. Following consultations among members of the Security Council, including the five permanent members, a mandatory resolution ordering both Iran and Iraq to end at once all military operations on land, at sea, and in the air was adopted unanimously on July 20, 1987. Resolution 598, under Articles 39 and 40 of the Charter of the United Nations, stipulates that the Security Council "demands that, as a first step toward a negotiated settlement, Iran and Iraq observe an immediate cease-fire, discontinue all military actions on land, at sea and in the air, and withdraw all forces to the internationally recognized boundaries without delay."[1]

Since the Korean War, no concerted action by the United Nations has been taken, as in this resolution, to end a war that lasted longer than either of the two world wars. "The resolution," as George Shultz, the American Secretary of State, remarked following the adoption of the resolution, "is scrupulously evenhanded." It is considered in the interests of all parties concerned and reflects the desire of the majority of the people in the region to end the war, which has taken an extraordinary toll in human life. It expresses the purposes of the United Nations Charter to maintain peace and order in compliance with International Law and to pursue settlement of disputes by peaceful means promptly. Almost simultaneously, both Iran and Iraq replied indicating their reactions to the Security Council resolution. While Iraq was prepared to accept the resolution, Iran, although neither accepting nor rejecting it, criticized its content. In his letter to Secretary General Perez de Cuellar of the United Nations, dated July 23, 1987, Tariq ʿAziz, Deputy Prime Minister and Foreign Minister of Iraq, stated, "The Iraqi Government welcomes the resolution and is ready to cooperate with you and with the Security Council so as to implement it in good faith with a view to finding a comprehensive, just, lasting and honorable settlement of the conflict with Iran."[2]

Iraq, however, made it clear that its acceptance of the cease-fire would depend on Iran's readiness to accept it in good faith as an "integral and indivisible whole" without conditions. ʿAziz also stated that the Security Council's reference to the withdrawal of all forces "without delay" was taken by the Iraqi government to mean that "the withdrawal shall be completed within a period not exceeding 10 days from the date of the general ceasefire." This period, according to Iraq, was based on the precedent set by the withdrawal of the Iraqi forces from Iranian territory on June 20, 1982. In accepting the Security Council resolution, the Iraqi government expressed its readiness to cooperate with the Secretary General's "mediation efforts" to achieve a settlement of the conflict with Iran in accordance with the principles of the United Nations Charter.

Before it replied in detail to Secretary Council Resolution 598, Iran sent two preliminary letters, signed by Foreign Minister ʿAli Akbar Velayati. In the first letter (July 22, 1987), addressed to the Secretary-General and delivered by Iran's permanent representative to the United Nations on July 24, 1987, it was stated that the "resolution . . . will be carefully considered and the position of the Islamic Republic of Iran will be declared in detail." In that letter, the Security Council was warned by Iran that Resolution 598, passed under American pressure, should not be used as a provocative measure by other members against Iran. The Security Council, the letter added, should call on the United States to halt its "military presence" in the Gulf and to withdraw its support for Kuwayt in exporting its oil for the benefit of Iraq.[3] In the second letter (August 10, 1987), Velayati complained that forty-two days had passed since the city of Sardasht had come under "Iraq's chemical bombardment." The attack, the letter stated, was immediately reported to the Security Council, but no move was taken to condemn Iraq. On what grounds, Velayati asked, is the "silence and inaction" of the United Nations justified? The answer to this and other matters, his letter added, will contribute to "our evaluation of the role of the United Nations Security Council vis-à-vis the war and its various dimensions."[4]

In another letter, dated August 11, 1987, Velayati set forth reservations to the Security Council resolution and "the conditions to be taken into consideration" were Iran to accept the resolution. Resolution 598, the letter stated, had been drawn up without prior consultation with the Iranian government and reflects "the Iraqi formulae for the resolution of the conflict." Moreover, it had been adopted under American pressure in favor of Iraq and its supporters in the war; it cannot, therefore, possibly be considered "balanced, impartial, comprehensive and practical." Nor has the Security Council, the latter added, even tried to determine which nation was the "aggressor" before it turned to end the war as "a breach of peace, thus necessitating recourse to Article 39 of the Charter." The Security Council, according to Iran, has failed to consider Iraq's "initiation of the war" on September 22, 1980, as a breach of the peace. In so doing, it has virtually "turned itself into a party to the conflict . . . and will not be able to play a positive and constructive role."[5] Nor is this all. In his letter, Velayati warned that the adoption of the resolution would exacerbate tension in the Gulf. The American actions, he added, "constitute clear violation of paragraph 5 of the resolution," which calls for restraint from any act leading to the further escalation of the conflict. Iraq, according to Velayati, violated the resolution by launching several offensive strikes immediately after its adoption. For this reason, Iraq was branded as an "aggressor" and, according to Iran, should be held responsible for the war and the payment of reparations as an essential element for the final resolution of the conflict. "Clear pronouncement on the responsibility of Iraq for the conflict," Velayati reiterated, "constitutes the most important element in the just resolution of the conflict." The eight-point plan of Perez de Cuellar, announced in March 1985, was referred to as the only practical plan dealing with the war, and it is "still a suitable ground for future efforts of the Secretary-General."

Although Velayati did not categorically reject the resolution, he made it quite clear that Iran's conditions must be met before the resolution would be accepted. In his letter, he stated, "Iran, as the victim of aggression, is the main party to determine how the war can be terminated, and no change can be effected in the course of the war as long as the conditions of the Islamic Republic of Iran are not met."[6]

Since only Iraq has accepted the Resolution and Iran, although not formally rejecting it, induced some members of the Security Council to argue against taking actions to apply sanctions against Iran, hoping that Iran might still be persuaded to accept a cease-fire without pressure. Other members noted the absence in Iran's reply of a demand to change the Iraqi regime as a sign of flexibility for further compromises. Such thoughts prompted Foreign Minister 'Aziz of Iraq to submit a letter, dated August 17, 1987, to Perez de Cuellar objecting to any proposal that might alter the text of the Security Council resolution and reaffirming his country's stand toward it. He went so far as to declare that Iran's response (August 11, 1987) to the resolution "constitutes a rejection." It represents, he added, "a form of selective approach" to its text, aiming obviously to continue the war and frustrate any concerted efforts to stop it. He quoted the final paragraph of a speech made on August 14 by 'Ali Khamini'i, President of Iran and Chairman of the Supreme Defence Council, in which he revealed his country's intention to continue the war by urging the people to be "prepared and . . . ready on the field of battle and at any place where war and mobilization officials want them to be in action." He, therefore, appealed to the Security Council to reject Iran's selective approach and to carry out the resolution "in letter and in spirit."[7]

Iran is not expected to accept the Security Council resolution without procrastinating and maneuvering to postpone action. Indeed, the Iranian regime maintained that it was winning the war; it thus wondered why the Security Council, "exactly at the time when [the war] is approaching its final stages," passed a resolution calling attention to the "breach of peace" and "necessitating recourse to Article 39 of the Charter." Iran's obvious desire not to stop the war prompted Iraq to resume the attack on tankers six weeks after Resolution 598 was issued. In its action, Iraq sought to put pressure on the Iranian regime to accept Resolution 598. Meanwhile, Perez de Cuellar visited Tihran and Baghdad in September for "consultation with Iran and Iraq and other states of the region," in order to inform the Security Council "to consider further steps to insure compliance" with it. Indeed, both countries, weary of the suffering and sacrifices in life and property, could not afford to turn down the Secretary General's offer to visit their capitals. Nor have their representatives to the United Nations been idle in contacting the Security Council directly. On his return, however, Perez de Cuellar reported to the Security Council that the Iranian leadership, taking a unified position, "indicated that Iran would silently honor a cease-fire but would announce its acceptance only 'at the conclusion of the Commission's work' as to which country was the aggressor." Such a stance obviously would reserve the option to Iran to resume the fighting by rejecting the commission's finding as flawed. Iraq had already made known its position that unless Resolution 598 was accepted unconditionally by Iran, it would not accept it. Since it had been announced that President Khamiʾini was to address the General Assembly on September 22, 1987, the Security Council, perhaps expecting that he would throw further light on Iran's stand toward Resolution 598, delayed its discussion on the implementation of the resolution.

Khamiʾini's address, however, proved very disappointing, as it turned out to be a lecture on the principles of the Iranian Revolution and a condemnation of "world domination" by the major powers; it contained virtually nothing constructive about Resolution 598. Khamiʾini denounced in particular the United States and its policy of intervention, which not only aggravated tension in the Gulf, but also tended to escalate and prolong the war by supporting Iraq. He criticized the Security Council for its failure, under pressure, to condemn Iraq's aggression against Iran when its first resolution, calling for a cease-fire and withdrawal to international frontiers, was issued on September 22, 1980. "Was Resolution 598," he pleaded, "only adopted to put pressure on the Islamic Republic of Iran?" "The United Nations," he went on to argue, "has an obligation, according to the very first Article of its Charter, to secure justice through the special process of taking measures against acts of aggression." He proposed, accordingly, to follow the precedent of the Nuremberg trials, maintaining that it has provided over forty years of peace in Europe. But he made no hint that Iran was prepared to accept a cease-fire or withdraw its forces to its frontier before the commission determined that Iraq had started the war, for which reparations should be paid.[8]

In conversations with Secretary General Perez de Cuellar the Iranian authorities have only hinted that they would accept an informal cease-fire once the commission formally stated its work, but no commitment was made to accept Resolution 598 if the finding of the commission was unacceptable to Iran. The question of which country started the war has become the basis on which Iran will demand punishing the Iraqi leaders and charging the regime over which they presided.

Iraq rejected Iran's accusation that it had started the war, although it agreed to set up a committee to determine guilt. Indeed, it had already proposed to refer the matter to the International Court of Justice for an advisory opinion. Since Iran gave no reply by the end of July that it had accepted Resolution 598 without conditions, Iraq resumed the tanker war. Likewise, Iran began to shell Iraqi installations and attack Kuwayti tankers escorted by American warships. Very soon, the tanker war escalated, and Iran attacked Kuwayti and Saudi tankers and installations, presumably in retaliation for Iraqi attacks on tankers carrying Iranian oil.

The American forces, whose primary task was to ensure free navigation in the Gulf in accordance with International Law, could not escape Iranian attacks. On September 21, the day before President Khami'ini was to deliver his address to the General Assembly, an Iranian navy ship, *Iran Ajr,* was laying mines. The American forces were bound to capture the mine-laying boat, as it was within sight of the American command and its act was an aggression on free navigation in international waters. The boat was accordingly destroyed and its surviving crew members were returned to Iran. On October 19, three or four Iranian gunboats fired at an American observation helicopter patrolling near Farsi Island. Several American helicopter gunships, in response to a call for help, opened fire on the Iranian boats and sank three of them. Six Iranians were rescued, but two later died. No less serious was the attack by American destroyers on three oil platforms, situated 120 miles east of Bahrayn, used for military operations by Iran inside the Gulf. Their destruction, described as "measured and appropriate," was in retaliation for the Iranian Silkworm missile attack a week earlier on an American-flagged tanker in Kuwayti waters. The Iranians were given twenty minutes' warning to flee the platform. This retaliation was further evidence to demonstrate to the Iranian authorities that the use of force is futile and that the acceptance of peaceful means for settlement of disputes is in the interest of all parties concerned.

Negotiations for the acceptance of Resolution 598 will probably be a long and tedious process, since no regime asserting such extremist doctrines as Iran is prepared to accept compromises without pressure. The Gulf War, after more than seven years, has proved to be unwinnable. Neither side has been able to win a decisive victory over the other. Nor are the regional or the global powers prepared to allow either side to be overrun by the other. Iran, however, maintains that time is on its side. By its maneuvers to delay the implementation of Resolution 598, it would be able to launch a series of surprise attacks on Iraqi territory, especially in the Basra area, in order to bring down the Ba'th regime, before it accepts Resolution 598 and enters into negotiations for a peaceful settlement with Iraq. But if Iran ever won the war, its victory would run contrary to Western policy that it should not win the war, as all other Arab regimes would be in jeopardy. Indeed, the influence of the Western powers in the region as a whole would be undermined, and the Security Council's balanced and moderate approach to the settlement of disputes frustrated. Pressure by means of an arms embargo on Iran, not to speak of economic sanctions, would be absolutely necessary to make it crystal clear that neither side shall be a winner and that settlement by peaceful means is the only way to end the war.

But should the fighting stop, will peace and good-neighborly relations between Iran and Iraq be reestablished? It is the message of this study that the Gulf War might come to an end at any moment, but unless the issues that led to war are resolved, resort to force is likely to recur at any time. War in the Gulf will come to an end only by a settlement acceptable to all the parties concerned. Quintus Ennius once said, Qui vicit non est victor nici victus fatetur. Without the consent of all parties, victory by one side would be meaningless and unjust, since peace can be broken by the other for flimsy reasons.

# Notes

## II. Persia's Rise as a Shi'i State and Its Conflict with the Ottoman Empire

1. A group in favor of 'Ali's candidacy to the Caliphate appeared soon after the Prophet Muhammad's death in 632. Tradition has it that he was supported by Ibn 'Abbas, a cousin, and a few others, but he was considered too young for the office. He was again nominated in 644, but the office went to 'Uthman. He was finally elected Caliph in 656, and became the first Imam in Shi'i traditions.

2. For the life of Shah Isma'il and his religious and military exploits, see Ghulam Sarwar, *History of Shah Isma'il Safawi* (Aligarh, 1939); E.G. Browne, *Literary History of Persia* (Cambridge, 1930), vol. 4, pp. 49–82.

3. For the accretion of rituals into Shi'ism under Safawi rule, see 'Ali al-Wardi, *Lamahat Ijtima'iya Min Tarikh al-Iraq al-Hadith* [Social Aspects of Modern Iraq History] (Baghdad, 1969), vol. 1, Chaps. 2 and 5.

4. The Ottoman sultans, in an effort to enhance their prestige in Islamic lands, assumed the title of caliph, claiming that after his conquest of Egypt in 1517 Sultan Salim had taken over the caliphial authority from the last 'Abhasid caliph then in exile in Cairo. Although a formal transfer of the title may not have taken place, Ottoman writers claimed that after his return to Istanbul, accompanied by the caliph, he proclaimed himself caliph of the Islamic world. See T. H. Arnold, *The Caliphate* (Oxford, 1924). pp. 139–58; Sati' al Husri, *al-Bilad al-Arabiya wa al-Dawla al-'Uthmaniya* [The Arab Countries and the Ottoman State] (Bayrut, 1960), pp. 42–46.

5. For the history of Perso-Ottoman relations and the military campaigns for control over the Arab provinces of Iraq, see S. H. Longrigg, *Four Centuries of Modern Iraq* (Oxford, 1925); 'Abbas al-'Azzawi, *al-Iraq Bayn Ihtilalayn* [Iraq Between Two Occupations] (Baghdad, 1935–56).

6. The principle *Cuius regio, eius religio*, first adopted at the Peace of Augsburg in 1555, became the basis of the European state system after the Treaty of Westphalia and provided the ground for the coordination of first the Christian states of Europe, and later states of different faiths throughout the world, into a community of nations.

7. For Nadir Shah's attempt to reconcile the Shi'i-Sunni creedal differences and

175

the compromise formula agreed upon at a conference held in Najaf in 1743, see Lockhart, *Nadir Shah* (London, 1938), pp. 100–101, 120–21, 210–11; 'Ali al-Wardi, *Social Aspects of Modern Iraq* vol. 1, pp. 131–34.

8. For Nadir Shah's proposal in the Treaty of 1746 (Article 2) to establish permanent missions, see Lockhart, *Nadir Shah,* p. 255.

## III. The Iraq-Iran Conflict during the Inter-War Years: The Confessional Aspect

1. Moreover, a few Shi'i followers of Arab descent adopted Persian citizenship in order to escape conscription and service in the Ottoman Army.

2. For an account of British relations with the Shi'i community, see Gertrude Bell, *Review of the Civil Administration of Mesopotamia* (London, 1920), pp. 27–33.

3. The proposals of the Cairo Conference were based on the viewpoint of the Foreign Office school of thought, which T. E. Lawrence, representing that school, had recommended to Colonial Secretary Winston Churchill (T. E. Lawrence, *Letters* [London, 1938], pp. 328–29). There were, however, other viewpoints representing local influence in Iraq (Philip Graves, *Life of Sir Percy Cox* [London, 1941], chap. 21).

4. For the Shi'i claim that the revolt of 1920 had influenced Britain in its decision to establish an Arab government, see Fariq al-Muzhir al-Fir'awn, *Al-Haqa'iq al-Nasi'a Fi al-Thawra al-Iraqiya, 1920* [The Plain Truth about the Iraqi Revolt of 1920], 2 vols. (Baghdad, 1952). The author is one of the tribal leaders who participated in the revolt. For Shi'i support of the candidacy of Faysal and the dispatch of Shaykh Rida al-Shibibi to Makka to request the Sharif Husayn's approval of Faysal's candidacy, see Shibibi's own account of his mission in ibid., pp. 568–77.

5. For an account of Curzon's policy in Persia, see Harold Nicolson, *Curzon: The Last Phase* (London, 1934), chap. 5. For a discussion of how Britain was able to perpetuate its influence in Iraq, see Daniel Silverfarb, *Britain's Informal Empire in the Middle East: A Case Study of Iraq, 1929–1941* (New York, 1986).

6. All vital statistics, whether before the First World War or during the interwar years, were based largely on estimates until the Iraqi government began to provide an official census after the Second World War. The population of the three provinces of Mawsil, Baghdad, and Basra under Ottoman rule was estimated at roughly 1.25 million in the mid-nineteenth century, and it was estimated shortly after the country passed under British control at about 2.5 million (Muhammad Hasan Salman, *Development of the Iraqi Economy* [Bayrut, 1951], vol. 1, pp. 39 and 40).

7. For an account of Faysal's conversations with Khalisi and other Shi'i leaders, see 'Ali al-Wardi, *Social History of Iraq,* vol. 6 (Baghdad, 1976), pp. 71–73, 107–11 (information based in part on the unpublished memoirs of Khalisi's son).

8. See the conversations of the King with Amin al-Rayhani about his dissatisfaction with the draft treaty and the country's need for British assistance against Turkish claims to recover the Mawsil Province, which the British forces had captured after

the armistice was signed in 1918 (Amin Rayhani, *Muluk al-Arab* [Arab Kings] (Bayrut, 1929), vol. 2, pp. 275ff.

9. For the rumor that the Ikhwan raid was inspired by the British, see ʿAbd al-Amir Hadi al-Akkam, *Nationalist Movement in Iraq, 1921–1933* (Baghdad, 1975), pp. 106–107; Colonial Office (Britain), *Report on Iraq, 1922–23* (London, 1924), pp. 4–5.

10. For the statement of the Secretary of State, see *Report on Iraq, 1922–23*, pp. 186–87.

11. See ibid., pp. 9–30; Colonial Office (Britain), *Report . . . on the Administration of Iraq, 1923–24* (London, 1925), pp. 9–13; A. J. Toynbee, *Survey of International Affairs, 1925* (London, 1927), vol. 1, pp. 531–34; Ali al-Wardi, *Social History of Iraq*, vol. 6, pp. 201ff.

12. The name of King Husayn, who had just been proclaimed Caliph, was substituted in the Friday sermon in place of the former Ottoman Caliph by a joint action of Sunni and Shiʿi mujtahids.

13. Colonial Office (Britain), *Report on the Administration of Iraq for the Year 1926* (London, n.d.), p. 25. This situation was reiterated in *Report by His Britannic Majesty's Government to the Council of the League of Nations on the Administration of Iraq for the Year 1927* [London, n.d.], p. 61.

14. League of Nations, *Minutes of the Permanent Mandates Commission,* 14th sess., November 1928, p. 177.

15. Ibid., p. 177.

16. Ibid.

17. Ibid., p. 178.

18. See Colonial Office (Britain), *Report on the Administration of Iraq for the Year 1928* (London, 1928), pp. 39–40.

19. League of Nations, *Minutes of the Permanent Mandates Commission,* p. 178.

20. Colonial Office (Britain), *Special Report on the Administration of Iraq for the Year 1930* (London, 1930), pp. 35–36.

21. For Persia's recognition of Iraq and the exchange of telegrams between the Shah of Persia and the King of Iraq, see *Report . . . on the Administration of Iraq for the Year 1929* [London, n.d.], pp. 36–39. For the Arabic texts of the telegrams and the Rustum Haydar mission to Persia, see ʿAbd al-Razzaq al-Hasani, *Tarikh al-Wazarat al-Iraqiya,* 4th ed. (Baghdad, 1974), vol. 2, pp. 207–14.

## IV. The Iraq-Iran Conflict during the Inter-War Years: The Diplomatic and Legal Aspects

1. For the text of the Iraq Nationality Law of 1924, see Colonial Office (Britain), *Report on the Administration of Iraq for 1925* (London, 1925), pp. 162–65 (hereafter referred to as *Iraq Report*).

2. The Pizder tribes of Sulaymaniya, in the central land frontier, were another group whose frontier crossing caused misunderstandings in the mid-1920s. (See *Iraq Report for 1926,* p. 26; *Iraq Report for 1927,* p. 62.)

3. See *Special Report by His Majesty's Government . . . to the Council of the League of Nations on the Progress of Iraq, 1920–31* (London, n.d.), pp. 169–78.

4. See *Report by His Majesty's Government . . . on the Administration of Iraq, 1932* (London, n.d.), pp. 14–15.

5. For a summary of the incidents and violations across the frontier, see J. I. Rawi, *al-Hudud al-Dawliya Wa Mushkilat al Hudud al-Iraqiya al-Iraniya* [International Frontiers and Problems of the Iraq-Iran Frontiers] (Baghdad, 1975), pp. 366–77.

6. For the text of the letters, see League of Nations, *Official Journal,* February 1935, pp. 196–97.

7. See M. Khadduri, *The Islamic Law of Nations: Shaybani's Siyar,* (Baltimore, 1965), pp. 11–14.

8. Ibid., pp. 17–18, 142–57. For the text of the Zuhab Treaty, see J. C. Hurewitz, ed., *The Middle East and North Africa in World Politics: A Documentary Record* (New Haven, Conn., 1975), vol. 1, pp. 25–28.

9. The Najaf Conference was attended by Shiʿi and Sunni scholars from Persia and Iraq, but Sunni scholars from other Islamic lands who seem to have had reservations about Nadir Shah's proposals were not invited. For an account of the Najaf Conference, see ʿAli al-Wardi, *Social History of Iraq* (Baghdad, 1969), vol 1, pp. 131ff.

10. For the text of the first Arzurum Treaty, see E. Hertslet, *Treaties . . . Between Great Britain and Persia* (London, 1891), pp. 163–68; Hurewitz, *Middle East and North Africa in World Politics,* pp. 219–21. For the text of the second Arzurum Treaty, see League of nations, *Official Journal,* January 1935, pp. 11–13; United Nations, *Treaty Series* (July 1969), pp. 1–3.

11. For a discussion of the Arzurum Treaty and the supplementary protocols, see Rawi, *al-Hudud al-Dawliya,* pp. 226ff; Khalid al-Izzi, *The Shatt al-Arab Dispute* (London and Baghdad, 1981), pp. 25–36.

12. See League of Nations, *Official Journal,* February 1935, p. 114.

13. Ibid., pp. 113–14.

14. Ibid., p. 116.

15. Ibid., p. 118.

16. Ibid., pp. 118–19.

17. For the background of the Saʿdabad Pact, see Arnold J. Toynbee, *Survey of International Affairs, 1936* (London, 1937), pp. 801–03; Abbas Khalatbary, *L'Iran et la pacte orientale* (Paris, 1938).

18. For texts of the Treaty of 1937 and the protocol, see League of Nations, *Treaty Series* 190, nos. 4401–4403; (1938): Appendix 1.

## V. The Revolutionary Movement in Iraq and Iran's Reaction to It

1. The story of how the Shah perceived his role and the steps taken to develop a "special relationship" between the United States and Iran has been told in a number of studies on Iran's foreign policy since the Second World War. For the most

important recent studies on Iran's foreign relations, see R. K. Ramazini, *Iran's Foreign Policy, 1941–1973* (Charlottesville, Va., 1975), and *The United States and Iran* (New York, 1982); Shahram Chubin and Sepehr Zabih, *The Foreign Relations of Iran* (Berkeley and London, 1974).

2. According to Trevelyan, British Ambassador to Iraq, Britain offered to sell arms to Qasim, but he preferred to obtain them from the Soviet Union, presumably because the Soviet Union offered a lower price (Humphrey Trevelyan, *The Middle East in Revolution* [London, 1970], pp. 152–55).

3. For the Kurdish War during the Qasim regime, see Edmund Ghareeb, *The Kurdish Question in Iraq* (Syracuse, N.Y., 1981), pp. 37–44; Majid Khadduri, *Republican Iraq* (London, 1969), pp. 173–81.

4. As evidence of Iranian complicity in the abortive coup, the Iraqi government made public intelligence information indicating that the third secretary of the Iranian embassy in Baghdad had first contacted the plotters on April 15, 1969, and some met with him at the Iranian embassy in Kuwait on September 28, 1969. Following these contacts, the Iranian government supplied the plotters with 3,000 submachine guns, 650,000 rounds of ammunition, and two mobile radio transmitters. It was also stated that secret correspondence between General Rawi and his collaborators had been transmitted through the Iranian embassy in Baghdad. Some of these letters and other documents from the plotters, including tape recordings of conversations between the plotters and Iranian representatives, seem to have been seized by the Iraqi government, as photocopies were made public in the press (Majid Khadduri, *Socialist Iraq,* [Washington, D.C., 1978], pp. 53–56).

5. For a discussion of whether the Shatt al-Arab is a national or an international river, see Jabir Ibrahim al-Rawi, *Al-Hudud al-Dawliya wa Mushkilat al-Hudud al-Iraqiya—al-Iraniya,* [International Rivers and the Problem of Iraq-Iran Boundaries] (Baghdad, 1975), pp. 327–31.

6. The accord, based on the peaceful methods provided under the League of Nations Covenant and other relevant treaties, was attached to the 1937 treaty shortly after it was concluded (Ibid., chap. 6).

7. In conversations with Abd al-Husayn al-Qatifi, former dean of the Baghdad Law College (who acted as an adviser to the Iraqi Foreign Office on the Shatt al-Arab dispute), I learned that Iran's abandonment of the resort to judicial settlement was prompted by the advice given to Iran by a number of lawyers, who thought that a judicial settlement on the basis of the 1937 Treaty would be unfavorable to Iran (Qatifi, interview with author, Baghdad, 1976 and 1985).

8. See Ministry of Foreign Affairs (Iraq), *al-I'tida'at al-Farisiya 'Ala al-Hudud al-Sharqiya Li al-Watan al-Arabi* [Persian Aggressions on the Eastern Borders of the Arab Homeland] (Baghdad, 1981), pp. 54–55.

9. The Iranian note, dated April 29, 1969, was delivered to the Iraqi government through the Iranian embassy in Baghdad. For background on the negotiations leading to the denunciation of the Treaty of 1937, see Rawi, *Al-Hudud al-Dawliya,* chap. 8; Ministry of Foreign Affairs (Iran), *Some Facts Concerning the Dispute Between Iran and Iraq Over Shatt al-Arab* (Tihran, 1969), pp. 78–79.

10. A navigation convention for the maintenance and supervision of the Shatt al-

Arab was expected to be laid down within a year from the coming into force of the 1937 treaty, in accordance with Article 5 of this treaty and Clause 2 of its protocol. If such a convention were not concluded within one year, then "this period may be extended by common agreement" of the parties concerned. Iraq, in accordance with the protocol, "shall undertake on the basis now in force all matters which are to be dealt with in this Convention" (Clause 2 [5]). Since the protocol is silent in the event no convention was concluded, Iraq maintained that responsibility for maintenance and supervision would devolve on it in order to continue its duties as before. Indeed, Iraq had called the attention of Iran to such a situation even before the 1937 treaty was ratified, but Iran failed to heed the Iraqi warning, although later it complained abbout Iraq's failure to deal with it. For Iraq's position on this situation, see Abd al-Husayn al-Qatifi, "Some Legal Aspects of the Termination of the Iraq-Iran Treaty of 1937," *Majallat al-Ulum al-Qanuniya* [Journal of Legal Studies] 1 (1969): 39–40, n. 39.

11. See W. E. Hall, *A Treatise on International Law,* 8th ed., ed. A. P. Higgins (Oxford, 1924), pp. 404ff; L. Oppenheim, *International Law,* ed. Lauterpacht (London, 1955), vol. 1, pp. 938–42; Lord Arnold Duncan McNair, *The Law of Treaties* (Oxford, 1961), p. 510. For a more recent discussion of the doctrine of *rebus sic stantibus,* see O. L. Lissitzyn, "Treaties and Changed Circumstances" *(Rebus Sic Stantibus), American Journal of International Law* 61 (1967): 895–945.

12. J. L. Brierly, *The Law of Nations,* 6th ed., ed. Humphrey Waldock (Oxford, 1963), p. 335.

13. For the text of the manifesto, see Khadduri, *Socialist Iraq,* pp. 231–40.

14. See Jalal Talabani, *Kurdistan Wa al-Haraka al-Qawmiya al-Kurdiya* [Kurdistan and the Kurdish Nationalist Movement] (Bayrut, 1971), pp. 321–23; Ghareeb, *Kurdish Question in Iraq,* pp. 61–62.

15. The Treaty of Sèvres (1920), concluded between the Ottoman sultan and the Allies, was repudiated by the Kamalist government and replaced by the Treaty of Lausanne (1923), which recognized no Kurdish state.

16. Other groups turned against Mulla Mustafa when the war broke out, such as the Kurdish Revolutionary party, led by Abd al-Sattar Tahir Sharif, and the Kurdish Progressive Group, led by Abd-Allah Isma il. The Kurdish Communist leaders, already participating in the National Front, also sided with the government (Khadduri, *Socialist Iraq,* pp. 91–95).

17. See Ghareeb, *Kurdish Question in Iraq,* pp. 142–45.

18. In his *White House Years,* Henry Kissinger states that President Nixon agreed "to encourage the Shah in supporting the autonomy of the Kurds in Iraq," presumably at Kissinger's suggestion; but he failed to relate the story of the embarrassing American involvement in the Kurdish War, although he promised to explain the American involvement in the second volume of his memoirs *(White House Years* [Boston, 1979], pp. 1204–65). For a detailed account of the American involvement in Kurdish affairs, based partly on the congressional Pike report and partly on interviews, see Ghareeb, *Kurdish Question in Iraq,* chap. 7.

19. After he stopped assistance to the Kurds, the Shah claimed that he entered

into an agreement with Iraq because the Kurds had no chance of winning the war. In a statement to the press, he said that he had ceased support for the Kurds because "they were making no progress in the war" (Joseph Kraft, "What Restrains the Shah", *Washington Post,* 27 April 1975). For the Iraqi side of the restraints—the increasing number of casualties and calls for a political settlement from among members of the Baʿth party—see Baʿth party, *al-Taqrir al-Markazi Li al-Muʾtamar al-Qatri al-Tasiʿ* [The Central Report of the Ninth Regional Congress, June 1982] (Baghdad, 1983), pp. 56–64.

## VI. A Short-lived Compromise: The Treaty of 1975

1. In 1973 and 1974, there were two direct contacts between Iranian and Iraqi representatives (although no official news about them was disclosed), but no common grounds for agreement were arrived at. The first took place in Baghdad when ʿAbbas Masʿudi, Vice-President of the Iranian Senate and editor of *Ittilaʿat,* arrived in Baghdad and met with several Iraqi leaders. The second took place in Geneva in 1974 when ʿAbbas Khalʿatbari, Foreign Minister of Iran, met with Murtada al-Hadithi, Foreign Minister of Iraq (Tariq ʿAziz, Foreign Minister of Iraq, interview with author, Baghdad, May 30, 1986). During my visit to Baghdad in 1974, Saddam Husayn, then Vice-President of Iraq, told me that Iraq was not prepared to accept a division of sovereignty over Shatt al-Arab (Saddam Husayn, interview with author, Baghdad, August 6, 1974).

2. The Shah spoke in French, and Saddam Husayn spoke in Arabic. Bumidian, fluent in both French and Arabic, translated from one language to the other for his two guests.

3. Muhammad Reza Pahlavi, *Answer to History* (New York, 1980), p. 133. The Shah's reference to "colonialist influences" is presumably to Britain, which had influence over both Iran and Iraq in the negotiations leading to the conclusion of the treaty of 1937.

4. Ibid.

5. For the full text, see Foreign Office (Iraq), *Documentary Dossier,* p. 285. See also Foreign Office (Iraq), *Iranian Aggression* (Baghdad, 1981), pp. 67–69.

6. See People's Democratic Republic of Algeria, *Memorandum Submitted by Algeria to the Conference of Sovereigns and Heads of State of OPEC Member Countries* (Algiers, 1975), p. 4.

7. Shortly after his return from Algiers, the Shah told Richard Helms, American ambassador to Iran, that his father had made two mistakes, which he had tried to correct. In his negotiations with Britain in 1933, he had not nationalized the oil industry. In his negotiations with Iraq for the Treaty of 1937, he had not insisted that the thalweg be the boundary line in the Shatt al-Arab. In the 1950s, said the Shah, he had accomplished nationalization, and at Algiers in 1975, he had established the thalweg as the border in the Shatt al-Arab (Richard Helms, interview with author, May 13, 1986).

8. For the texts of the treaty and the protocols, see Majid Khadduri, *Socialist Iraq* (Washington, D.C., 1978), Appendix E.

9. Since the United States had endorsed the Shah's promises of support to Mulla Mustafa, the Shah's decision to stop his assistance to the Kurds without prior American consultation must have been disappointing to American policy makers. On learning the news of the Algiers Agreement, Henry Kissinger is reported to have been upset with the Shah's single-handed decision to stop assistance to the Kurds.

10. In an interview with Mulla Mustafa and his counselor Muhsin Diseyi in Washington, on December 12, 1976, I asked them whether Kurrdish dependence on foreign powers had ever been helpful in resolving the Kurdish problem in Iraq. They replied that the Kurds had learned the lesson that they must depend on themselves, presumably that they would have to resolve their differences with Iraq by peaceful methods because resort to war had aggravated tensions and hurt the interests of both sides.

11. The three principal sectors that remained under Iranian control were Sayf Sa'd, Zayn al-Qaws, and Maymak (Foreign Office [Iraq], *The Iraq-Iran Conflict in International Law* [Baghdad, 1981], p. 58). These sectors are strategically important, as they overlook some of the Iraqi towns across the frontiers and are vulnerable to air attack by Iran. "This," says Sa'dun Hamadi, Foreign Minister of Iraq, "was exactly what happened when the Iranian forces attacked Iraq on September 4, 1980" (Sa'dun Hamadi, *Mulahazat Hawl Qadiyat al-Harb Ma'Iran* [Remarks About the Question of the War with Iran] [Baghdad, 1982], p. 88).

12. Ibid., p. 24.

13. Sa'dun Hamadi, interview with author, Baghdad, November 21, 1985.

14. See the speech by the head of the Iranian delegation before the United Nations General Assembly (United Nations, General Assembly, Official Records, A/35/PV33, October 10, 1980); the speech by Premier Muhammad 'Ali Raja'i before the Security council (United Nations, Security Council, Official Records, S/PV 2251, October 17, 1980). For other statements made by Foreign Minister Sadiq Qutb-Zada and other high officials, see the Islamic Republic (Iran), *The War of Aggression Against the Islamic Republic of Iran by the Iraqi Ba'th Regime* (Tihran, 1981).

15. Foreign Office (Iraq), *Iraq-Iran Conflict in International Law*, p. 60.

16. Ibid., p. 64.

## VII. Changes in the Regimes of Iran and Iraq and Political Developments Leading to the Gulf War

1. Khumayni, "Hukumat-i Islami" [Islamic Government], in *Islam and Revolution,* trans. Hamid Algar (Berkeley, 1981), pp. 27–149.

2. For Khuymayni's views on the Imamate, see ibid., p. 72.

3. For the text of Khymayni's lectures on jihad, "Mubaraza Ba Nafs Ya Jihad-i Akbar" [The Soul's Struggle or the Great Jihad], see *Islam and Revolution,* trans. Hamid Algar, pp. 351–60. For the meaning of the concept of the "greater jihad" in accordance with Sunni classical texts, on which Khumayni has drawn, see Majid

Khadduri, "The Great War," *Aramco World Magazine* (July–August 1968): 25–27.

4. According to Sunni texts, the concept of the jihad as a defensive-offensive war was stressed by Shafiʾi, founder of the school of law bearing his name, and it was accepted by most commentators from the ninth century to the fourteenth in the Muslim world (Shafiʾi, *Kitab al-Umm* [Cairo, 1904] vol. 4, pp. 84–85. For the viewpoint that the jihad was a defensive war, which became prevalent after the fall of the ʿAbbasid Caliphate in the thirteenth century, see Ibn Taymiya [d. 1327], *Majmuʿ Rasaʾil* [Collected Essays] (Cairo, 1949), pp. 115–46.

5. "Fight those who believe not in God and the Last Day and do not forbid what God and Apostle have forbidden—such men as practice not the religion of truth, being of those who have been given the Book—until they pay the tribute out of hand and have been humbled" (*Qurʾan* IX, 29).

6. For an exposition of this doctrine, see Ayat-Allah Ahmad Jannati, "Nizam-i difaʾ wa jihad dar Quran-i Karim," [Defence and Jihad in the Qurʾan], *A Quarterly Journal of Islamic Thought and Culture* 1 (1984): 39–54.

7. For commentary by classical scholars, see Muhamad Bin Jarir al-Tabari, *Tafsir,* ed. Mahmud Shakir (Cairo, 1958), vol. 14, pp. 574–76.

8. For the resolutions of the Arab summit at Baghdad, see Ministry of Information (Iraq), *Wathaʾiq Li Mujabahat al-Tahadi* [Documents on Opposition to Confrontation] (Baghdad, 1979); Saddam Husayn, *al-Iraq wa al-Siyasa al-Dawliya* [Iraq and World Politics] (Baghdad, 1981).

9. President Bakr, interview with author, Baghdad, August 14, 1974.

10. For the text of the unity pact, see *al-Jumhuriya* (Baghdad), 27 October 1978.

11. The foregoing account is based in part on conversations with several Iraqi and Syrian leaders during my visits to Baghdad and Damascus in the spring and fall of 1978. Above all, I have drawn on my conversations with Tariq ʿAziz, foreign minister of Iraq, during my visits to Baghdad in November 1985 and May 1986, and on conversations with several other Syrian and Iraqi informants, who preferred not to be identified. For the official Iraqi version of the unity negotiations, see Baʿth party, *al-Taqrir al-Markazi Li al-Muʾtamar al-Qatri al-Tasiʿ* [The Central Report of the Ninth Regional Congress, June 1983] (Baghdad, 1983), pp. 319–32.

12. For the text of Bakr's speech see *al-Thawra* (Baghdad), 17 July 1979.

13. For Samarraʾiʾs involvement in the so-called Kazzar plot, see Majid Khadduri, *Socialist Iraq* (Washington, D.C., 1978), p. 65.

14. During my visit to Damascus in 1981, I learned from conversations with several Syrian leaders, including President Asad, that tahe differences between Syria and Iraq were ideological—Pan-Arab ideals versus regional interests—and not personal (Majid Khadduri, *Arab Personalities In Politics* [Washington, D.C., 1981], p. 220).

15. Khadduri, *Socialist Iraq,* p. 76. For the tribute paid to Saddam Hysayn's leadership by his party at the Ninth Regional Command Congress in 1982, see Baʿth party, *al-Taqrir al-Markazi,* pp. 36–43.

## VIII. The Iraq–Iran War: The Diplomatic and Legal Aspects

1. See Musa al-Musawi (Musavi), *al-Thawra al-Ba'isa* [The Wretched Revolution] (Bayrut, 1980), pp. 98, 160. Musawi, who knows Khumayni, was a member of the Iranian Majlis (Parliament) during the 1960s.

2. For the text of the memorandum, see Foreign Office (Iraq), *The Iraq-Iran Conflict in International Law* (Baghdad, 1981), Appendix 1, pp. 115–16. See also Saddam Husayn's speech before the third summit of the Organization of the Islamic Conference at Ta'if, Saudi Arabia, 1981 (Baghdad, 1981), p. 28.

3. See the speech by Saʿdun Hamadi, Foreign Minister of Iraq, before the United Nations Security Council, October 15, 1980 (United Nations, Security Council, Official Records, S/PV 2250; Foreign Office [Iraq], *Documentary Dossier* [Baghdad, 1981], pp. 234–35). Hamadi said that he had sought clarification of the discrepancy between the two texts through diplomatic channels. He was told that the cable from the Iranian Foreign Office was the official one and that an investigation about the edited text of the other would be reported later. Nothing about the investigation or correction of the cable was ever made known to Iraq.

4. Musa all-Musawi, in a personal account of his mission to Iran in 1979, said that in a conversation with then Vice President Saddam Husayn, he had been told that the Baʿth government was quite prepared to open a new chapter in the relationship between the two countries. On his arrival in Tihran, Musawi visited Bazargan, Bani-Sadr, and Khumayni. Bazargan and Bani-Sadr told Musawi that they were prepared to settle all differences between the two countries by peaceful methods, but Bani-Sadr warned that Khumayni was opposed to negotiations with the Baʿth leaders. When Musawi went to talk with Khumayni at Qum, he found him quite adamant in his stand, although Musawi pleaded with him that the Baʿth party had, after all, helped him during his exile in Iraq (Musawi, *al-Thawra*, pp. 74–90).

5. For this and other statements, see Saʿdun Hamadi's speech (United Nations Records S/PV 2250; Foreign Office [Iraq], *Documentary Dossier*, pp. 238–39).

6. For a list of violations on land and in the air by Iran, see Foreign Office (Iraq), *Documentary Dossier*, pp. 28–49.

7. One of the suspects, ʿAbd al-Amir Hamid al-Ansari, admitted in the trial that he and his collaborators had received explosives and money from Iran to carry out subversive activities in Iraq. His confession, reiterated over Baghdad radio and television on April 24, 1980 was communicated to Secretary General Waldheim for circulation among the United Nations delegations (United Nations General Assembly, Official Records, A/35/68, May 28, 1980).

8. See Foreign Office (Iraq), *Documentary Dossier*, p. 194.

9. See Saddam Husayn's speech, p. 37; Foreign Office (Iraq), *Documentary Dossier*, p. 198.

10. See Saddam Husayn, *al-Iraq wa al-Siyasa al-Dawliya* [Iraq and World Politics] (Baghdad, 1981), pp. 275–76; for Hamadi's speech, see United Nations, General Assembly, Official Records, 35th Sess., 22nd plenary meeting, October 3, 1980; Foreign Office (Iraq), *Documentary Dossier*, p. 229.

11. For the text of the speech, see Foreign Office (Iraq), *Documentary Dossier,* p. 212.

12. Ibid., p. 200.

13. For the text of the note, see United Nations, Security Council, Official Records, S/14249, November 11, 1980.

14. For the text of the note to Iran (November 16, 1980) and the letter to the Secretary General of the United Nations (November 25, 1980), see United Nations, Security Council, Official Records, S/14272, November 26, 1980.

15. See Iraq's note of protest to Iran in Foreign Office (Iraq), *Documentary Dossier,* pp. 202–204; Saddam Husayn's speech, pp. 39–40.

16. See Foreign Office (Iraq), *Iraq-Iran Conflict in International Law,* pp. 24–27, 68–74, 77–89.

17. United Nations, General Assembly, Official Records, A35/483 and S/14191, September 24, 1980 (Annex), p. 3.

18. See United Nations, Security Council, Official Records, S/14193, September 23, 1980.

19. United Nations, Security Council, Official Records, S/14199, September 26, 1980 (Annex II).

20. United Nations, Security Council, Official Records, S/14190, September 28, 1980.

21. See President Husayn's letter to Secretary General Waldheim in United Nations, Security Council, Official Records, S/14203, September 29, 1980 (Annex); Foreign Office (Iraq), *Documentary Dossier,* p. 225.

22. For the text of Bani-Sadr's letter, see United Nations, Security Council, Official Records, S/14206, October 1, 1980.

23. *New York Times,* 4 October 1980.

24. In support of Khumayni and Bani-Sadr, 'Ali Akbar Rafsanjani made a statement in the Majlis (September 29, 1980) in which he declared that there was no question of a "cease-fire" between Iran and Iraq. *New York Times,* 30 September 1980.

25. For the text of Raja'i's speech, see United Nations, Security Council, Official Records, S/PV 2251, October 17, 1980.

26. For the text of Hamadi's speech, see United Nations, Security Council, Official Records, S/PV 2251, October 17, 1980.

27. United Nations, General Assembly, Official Records, 35th sess., October 1, 1980, p. 337.

28. In addition to Sekou-Touré, representing Gambia, and the Secretary General of the Islamic Conference Organization, the Good Offices Committee was composed of the representatives of Senegal, Pakistan, Bangladesh, Turkey, Guinea, Malaysia, and the Palestine Liberation Organization.

29. See *Kayhan International* (Tihran), 1 March 1981,

30. *Kayhan International,* 28 February 1981.

31. For Khumayni's speech, see *Kayhan International,* 2 March 1981, Appendix VI.

32. See *Kayhan International,* 5 March 1981.

33. Islamic Peace Committee, "Suggestions for a Package Peace Settlement" (Jidda, March 1981, Mimeographed).

34. *Kayhan International,* 5 March 1981.

35. *Kayhan International,* 7 March 1981.

36. Khumayni and other Iranian leaders had suggested a settlement on the basis of Islam. While visiting Washington in 1984, I learned in a conversation with the Secretary General of the Islamic Conference Organization that a peaceful settlement on the basis of Islamic principles might be acceptable to Iran and Iraq. When, however, the opinions of several Muslim scholars in Iran and Iraq were solicited on the subject, their replies were sharply conflicting. While the Iraqi scholars argued in favor of peace, the Iranian scholars held that the aggressor be punished by war (jihad), presumably both groups of scholars arguing on the basis of authoritative Islamic texts.

37. United Nations, Security Council, Official Records, S/14251, November 11, 1980; United Nations, Security Council, Official Records, S/15449, October 7, 1982. See also *New York Times,* 23 February 1981. Cf. K. de Young, "Swedish Arms Scandals Mar Peacemaking Image," *Washington Post,* Sept. 5, 1987, alleging that Palme was involved in secret arms sales on behalf of Swedish companies while on his peace mission in Iran and Iraq.

38. For the text of the speech, see Foreign Office (Iraq), *Documentary Dossier,* pp. 216–23.

39. Ibid; Saddam Husayn, *al-Iraq wa al-Siyasa al-Dawliya,* p. 252.

40. In a statement made to foreign correspondents, Taha Yasin Ramadan, First Deputy Premier, declared that Iraq would "continue to clean up the region and take the cities of Arabistan," that "the Arabistan oil will remain Iraqi as long as Tihran will not negotiate," and that the three islands at the Strait of Hurmuz were "Arab territory" (*New York Times,* 21 October 1980). For the argument that Iran had denied the national existence of the Arabs of al-Ahwaz and deprived them of their rights to a separate existence, see Ministry of Foreign Affairs (Iraq), Persian Violations of the Eastern Frontiers of the Arab Homeland (Baghdad, 1981), pp. 107–37.

41. *New York Times,* 18 October 1980.

42. See R. A. Falk, "International Law and the Peaceful Settlement of the Conflict," in *The Iraq-Iran War,* ed. A. E. H. Dessouki (Princeton, M. J.: Woodrow Wilson School of Public and International Affairs, 1981), p. 83.

## IX. The Iraq-Iran War: The Ideological Aspect

1. In Iran, as in other Islamic countries, there are differences of opinion on the relationship between civil and religious authorities. For example, Ayat-Allah Shariʿat-Madari held that the mujtahids, confining their duty to advise on religious and legal matters, should not be involved in politics; but Khumayni and his disciples insisted that political and religious matters are inseparable, and the mujtahids cannot close their eyes if rulers violate Islamic law and Islamic standards.

2. See Munazamat al-ʿAmal al-Islami, *Al-Iraq Bayn al-Tanzimat wa al-Khiyarat al-Matruha* [Iraq's [Choice] Between Existing Organizations and Proposed Options] (Iraq, n.d.), pp. 11–13, 50–54.

3. Since the ʿulama of Najaf (Iraq) were highly regarded in Iran, the cooperation between the Shah and the leading clerics in Iraq had, in effect, undermined the position of the Iranian mujtahids who were opposed to the Shah (Musa al-Musawi, *al-Thawra al-Baʾisa* [The Wretched Revoluttion] [Bayrut, 1980], pp. 84ff).

4. I have it on the authority of one of my informants that shortly after his arrival at Najaf, Khumayni paid a courtesy visit to al-Hakim. In the course of their conversation, Khumayni, an activist, asked al-Hakim whether he should not speak out against oppression and corruption in Iran. Al-Hakim, holding that the function of the mujtahid was confined to religious affairs, replied that it was none of his business to criticize the Iranian government. Disagreeing with al-Hakim, Khumayni is reported to have remarked that although both he and al-Hakim were descendants from ʿAli, the first Imam, the difference lies in the fact that al-Hakim was a great grandson of al-Hasan, eldest son of ʿAli, while he was a great grandson of al-Husayn (the implication being that al-Hasan, advocating quietude, submitted to the Sunni Caliph, but al-Husayn, defying authority, became the martyr who stood against tyranny and injustice).

5. See Musawi, *al-Thawra,* pp. 157–61; cf. Hamid Algar, *The Islamic Revolution* (Qum, 1981), p. 56.

6. It is said that Sadr was appointed as Khumayni's deputy in Iraq. See Saʿdun Hamadi, *Mulahazat Hawl Qadiyat al-Harb Maʿ Iran* [Remarks About the War with Iran] (Baghdad, 1982), p. 30; compare Daʿwa party, *Proclamation from the Islamic Daʿwa party to the Iraqi Nation,* Central Information, No. 8.

7. For the life of the elder Sadr, see Agha Bazrak al-Tihrani, *Aʿlam al-Shiʿa* (Najaf, 1954), vol. 1, pp. 445–449.

8. For Sadr's role in politics, see Majid Khadduri, *Independent Iraq,* 2d ed. (London, 1960), pp. 270–72.

9. For Sadr's role in the Daʿwa party and his followers, see Fadil al-Barrak, *Tahaluf al-Adhdad* [Coalition of Opposites] (Baghdad, 1985), pp. 85–101.

10. For a description of Shiʿi religious (educational) institutions, see Fadil Jamali, "Theological Colleges of Najaf," *Muslim World* 1 (1960): 15–22.

11. For the role of Iranian schools in Iraq and their impact on the Iran–Iraq relationships, see Fadil al-Barrak, *Al-Madaris al-Yahudiya wa al-Iraniya Fi al-Iraq* [Jewish and Iranian Schools in Iraq], 2d ed. (Baghdad, 1985), pp. 93ff.

12. Khumayni, "Hukumat-i Islami" [Islamic Government], in *Islam and Revolution,* trans. Hamid Algar (Berkeley, 1981), p. 47 and passim. See also Ministry of Culture and Information (Iraq), *Iran's Hostility to Iraq: Statements by Khumayni and His Aides* (Baghdad, 1984).

13. See publications of the Islamic Action Organization, an underground Islamic organization in Iraq.

14. See Daʿwa party, *Proclamation of an Understanding [among the ulama] from*

*the Islamic Da'wa Party to the Iraqi Nation,* No. 8, n.d. The proclamation was addressed to all the people of Iraq and to all political groups and organizations—Arabs, Kurds, Muslims, and non-Muslims—to unite against the regime.

15. Earlier, almost all the western media echoed the Iranian claim that Iraq was the aggressor. Godfry Jansen, the Levant correspondent of the London *Economist,* noted that owing to the "inept Iraqi propaganda, the media battle is one that the Iraqis lost from the very start" (Godfry Jansen, "The Gulf War: The Contest Continues," *Third World Quarterly* [October 1984]: 952).

16. See the texts of the address by Ayat-allah Khumayni to the Islamic Mediating Committee (March 1, 1981) and by President Husayn to the Iranian people (June 14, 1985), Appendixes VI and VII.

17. See the statement by Ali Shams Ardakhani, of the Iranian delegation to the United Nations, to the General Assembly (United Nations, General Assembly, Official Records, A/35/PV33, 1980).

18. Fadil al-Barrak, head of the Iraqi Security Department, interview with author, Baghdad, June 1, 1986.

19. The Ba'th position on the relationship between religion and politics has been fully explained by Saddam Husayn in a speech entitled, "Remarks About Religion and the Heritage" (Baghdad, 1977). See also Saddam Husayn, *Al-Iraq wa al-Siyasa al-Dawliya* [Iraq and World Politics] (Baghdad, 1981), pp. 277–78.

20. Today, Shaykh 'Ali Kashif al-Ghita (born in Najaf in 1904) is the leading Shi'i scholar in Iraq. He advocates the separation of religious from civil authorities, and holds that the mujtahids should confine their guidance to religious affairs. He published several works on Islamic law and religion. His most recent books are one on *Fiqh* (Islamic jurisprudence), published in 1979, and another on logic, published in 1982. For the role of the Shi'i ulama in Arab nationalist activities, see Wamid Nazmi, "The Shi'a of Iraq and Arab Nationalism," *al-Mustaqbal al-Arabi* 42 (1982): 74–100.

21. As an example of disagreement with Khumayni's creedal views (although this is not a typical example), twelve Shi'i ulama from Iraq representing various organizations signed a *fatwa* in which they denounced Khumayni's view that the "Prophet Muhammad was incapable of achieving all goals as he wished them." They objected that Khumayni's use of the word *incapable* is improperly applied to the Prophet. For the text of the *fatwa,* see *Al-Thawra* (Baghdad), 2 April 1987.

## X. Involvement of the Gulf States in the Iraq-Iran War

1. For the text of the Arab National Declaration, see Colin Legum et al., *Middle East Contemporary Survey* (New York and London, 1981), pp. 224–25; Nicola Firzli, ed., *The Iraq-Iran Conflict* (Paris, 1981), Appendix 1, pp. 167–70.

2. For Iraq's attitude toward the Gulf, see Edmund Ghareeb, "Iraq and the Gulf," in F. R. Axelgard, ed., *Iraq in Transition* (Boulder, Colo., and London, 1986), pp. 59–83.

3. For the viewpoint that Iraq was fighting an Arab war and defending the fron-

tiers of Arab lands, see ʿAbd al-Latif al-Hadithi *et al., Al-Hudud al-Sharqiya li al-Watan al-Arabi* [The Eastern Frontiers of the Arab Homeland] (Baghdad, 1981).

4. Saudi Arabia, Kuwayt, and Jordan offered their airfields as shelters for Iraqi planes as a token of support as soon as the war started.

5. For King Husayn's idea about the relationship between Islam and nationalism, see King Husayn, "How to Unite the Arab World," *Life International,* 23 May 1960, p. 30.

6. The name Yarmuk was a reminder of the battle fought by the Arabs in A.D. 636 in which they defeated the Byzantine army in the vally of Yarmuk in Jordan. The purpose of the volunteers was symbolic, to enhance the morale of the Iraqi army, as Jordan could not possibly commit a large part of its force outside the country because it was needed for defence of its western borders.

7. Former Jordanian Premier Mudar Badran; Premier Zayd al-Rifaʿi; Chief of the Royal Court and former Foreign Minister Marwan al-Qasim; Former Minister of Culture and Information ʿAdnan Abu ʿAwda; and several other men in high offices, interviews with author, Amman, Jordan, June 1987.

8. In Saudi Arabia, there have always been extremist groups opposed to change brought about under the impact of Western technological innovations, as evidenced by several protests and incidents culminating in the seizure of the Kaʿba Mosque at Makka in November 1979, under the inspiration of Khumayni's call for revolutionary activities.

9. For the text of the mobilization law as well as other laws concerning internal security, see Ministry of Information (Kuwayt), *Kuwayt Gazette* for 1980 and subsequent years.

10. On Qutbzada's arrival at Kuwayt, there was an attempt on his life, an act implying that on the popular level, Kuwayt was opposed to Iran's hostile actions against Iraq.

11. Mazhar A. Hameed, "After Mecca: Saudi Arabia Is More Stable than It Looks," *Washington Post,* 9 August 1987.

12. Sadat told me (in a conversation in Cairo on March 14, 1981) that he had decided to support Iraq when a number of leaders from Saudi Arabia and ʿUman inquired whether Egypt was prepared to sell them weapons intended for Iraq. He said that he had decided to offer it to Iraq himself. My conversation with Sadat was not for the purpose of this study (although in the course of the conversation the Gulf War was touched on), but for an essay on Sadat's leadership, "Anwar al-Sadat of Egypt," in Majid Khadduri, *Arab Personalities in Politics* (Washington, D.C., 1981), pp. 125–80.

13. For the rift between Iraq and the Arab Steadfast and Confrontation Front, see Tariq ʿAziz, *al-Siraʿ al-Iraqi-al-Irani* [The Iraq-Iran Struggle] (Bayrut, 1981), pp. 85–98; Baʿth party, *al-Taqrir al-Markazi Li al-Muʾtamar al-Qatri al-Tasiʿ* [The Central Report of the Ninth Regional Congress] (Baghdad, 1983), pp. 311–19.

14. Syria's interruption of the flow of oil through the pipeline is considered by Iraq a violation of the 1982 agreement. It has referred its dispute with Syria to the Arab Organization of Oil Exporting Countries (OAPEC) for an advisory opinion,

and both countries have accepted the jurisdiction of the OAPEC's judicial organization over the case. In a conversation in 1987 with Dr. ʿAbd al-Latif al-Shawwaf, a
member of the judicial panel, I asked him about the nature of the dispute. He said
that at the outset Syria considered the dispute political and therefore nonjudicial,
but its legal counsel agreed to discuss the merits of the case. Iraq, considering the
agreement still binding, agreed to accept OAPEC's advisory opinion. The pipeline
is obviously outside the zone of war and is a purely commercial agreement from
which Syria derives in the long run greater material benefit than it received from
Iran. Syria's political motivation to stop the flow of oil obviously has no bearing on
the validity of the 1982 pipeline agreement.

15. Nora Bustani, "Syria's Economic Crisis Seen Threatening Ties with Iran,"
*Washington Post,* 17 July 1987.

16. The Soviet Union, the ally of both Syria and Iraq, offered its good offices by
inviting the foreign ministers of the two countries to Moscow as early as 1981. This
offer was extended on more than occasion, to no avail.

17. Tariq ʿAziz, Deputy Premier and Foreign Minister of Iraq, interviews with
author, Baghdad, May 30, 1986, and April 18, 1987; Mudar Badran, former Premier
of Jordan, and Zayd al-Rifaʿi, Premier of Jordan, interviews with author, Amman,
Jordan, June 17 and 20, 1987. For President Asad's views on the Gulf War and his
relationships with the Iraqi leaders, see Ahmad Qurna, ed., *Hafiz al-Asad: Mawsuʿa
Kamila* [Collected Papers] (Aleppo and Bayrut, 1985), vol. 4, pp. 151–83; interview
with Hafiz al-Asad, editor of *al-Qabas* (Damascus), 24 January 1987.

18. For the viewpoint that Syria's support of Iran is a betrayal of Arabism, see
Saʿdun Hamadi, *Mulahazat Hawl Qadiyat al-Harb Maʿ Iran* [Remarks About the
War with Iran] (Baghdad, 1981), pp. 85–98.

19. For Qadhdhafi's involvement in the disappearance of al-Sadr, see Peter Ther-
oux, *The Strange Disappearance of Imam Moussa Sadr* (London, 1987).

20. Saudi Arabia's diplomatic relations with Libya were resumed in 1986, but
differences between the two countries have not yet been resolved.

21. For the background and settlement of the dispute, see Quincy Wright, "The
Mosul Dispute," *American Journal of International Law* 20 (1926): 453–63; Fadil
Husain, *Qadiyat al-Mawsil* [The Mosul Problem] (Baghdad, 1955).

22. Turkey's attendance at the Islamic summit in 1981 was its first participation
in an Islamic conference since the abolition of the Caliphate and the promulgation
of the Turkish constitution in 1924, in which the principle of the separation of state
and religion was adopted. Turkey made it clear, however, that its participation in
the Islamic conference was not intended to imply that the secular character of the
state had been altered.

23. Only a few members of the Barzani clan, who supported Mulla Mustafa's
struggle against the regime, were deported to Nasiriya, in the south of Iraq, and their
estate was confiscated.

24. Ismat Kittani head of the Iraqi mission to the United Nations, interview
with author, March 1987. Kittani is an Iraqi Kurd and an authority on Kurdish affairs.

25. The author's observation during his visit to the Mawsil Province in 1987.

## XI. In Search of Peace and Security in the Gulf

1. For a discussion of conditions in the Gulf and piracy as an expression of resentment against foreign intervention, see Sultan Muhammad al-Qasimi, *The Myth of Arab Piracy in the Gulf* (London, 1986); Muhammad M. Abdullah, *The United Arab Amirates* (London, 1978).

2. Persia recovered full sovereignty over its coastal area after the First World War. Iraq, one of the successor states to the Ottoman Empire, passed under the British Mandate before it became independent in 1932. The other Gulf principalities—Kuwayt, Bahrayn, Qatar, and the Trucial Coast (later the United Arab Amirates)—became independent after the Second World War. For British policy during the era before the Gulf states achieved independence, see Sir Arnold Wilson, *The Persian Gulf* (Oxford, 1928); John Marlow, *The Persian Gulf in the Twentieth Century* (London, 1962).

3. In the early 1970s, Iraq might have become a joint partner, but its preoccupation with domestic affairs, especially the Kurdish problem, limited its interest in Gulf affairs to the control of the Shatt al-Arab and the settlement of frontier problems. Its involvement in Gulf security began to develop later in the 1970s, following the Algiers Agreement (1975). For a discussion of the interdependence of the Arab Gulf countries in all matters of security, social and economic cooperation, and human and material resources, see Hasan al-Ibrahim, ''al-Khalij Wa al-Watan al-ʿArabi,'' *al-Mustaqbal al-Arabi* (Bayrut, 1984), vol. 7, pp. 4–9.

4. For Iran's annexation of the three islands, see Jasim M. Abdulghani, *Iraq and Iran: The Years of Crisis* (London, 1984), pp. 89–94.

5. For the aims of the Arab League and the events leading up to its establishment in 1945, see A. M. Gomaa, *The Foundation of the League of Arab States* (London, 1977).

6. For the legal position of the Arab League in its relationship with the United Nations, see Majid Khadduri, ''Arab League as a Regional Arrangement,'' *American Journal of International Law* 40 (1946): 756–77.

7. For the text of a conversation between V. M. Molotov, Chairman of the Council of People's Commissar and Commissar for Foreign Affairs of the U.S.S.R., and J. von Ribbentrop, the German foreign minister, in Berlin on November 12, 1940, see R. S. Sontag and J. S. Beddie, eds., *Nazi-Soviet Relations, 1939–1941* (Washington, 1948), p. 222.

8. For the text of the declaration, see J. C. Hurewitz, ed., *The Middle East and North Africa in World Politics: A Documentary Record* (New Haven, Conn., and London, 1979), vol. 2, pp. 680–84.

9. See Mohammad Reza Pahlavi, *Answer to History* (New York, 1980), p. 13.

10. See Cyrus Vance, *Hard Choices: Critical Years in America's Foreign Policy* (New York, 1983), pp. 388–89.

11. The Afghan wars, called in England ''the great game,'' resulted in the recognition of Afghanistan as not only an independent, but also a buffer state (see W. K. Fraser-Tytler, *Afghanistan: A Study of Political Development in Central Asia* [London, 1950], chaps. 7–9).

12. Zbigniew Brzezinski, *Power and Principle,* rev. ed. (New York, 1985), p. 444.

13. Ibid., p. 446.

14. In his State of the Union address, Carter said, "We are prepared to work with other countries in the region to shape a cooperative security framework that respects different values and political beliefs, yet which enhances the independence, security and prosperity of all."

15. Brzezinski, *Power and Principle,* p. 447.

16. For statements about American policy toward the Gulf, see U.S. Congress, House Committee on Foreign Affairs, *Hearings Before the Subcommittee on Europe and the Middle East of the Committee on Foreign Affairs and the Joint Economic Committee,* 97th Cong., 2nd sess., 10 May 1982 (Washington, D.C., 1983).

17. See Alexander Haig, *Caveat: Realism, Reagan, and Foreign Policy* (New York, 1984), chap. 9.

18. The cost of preparing the Gulf for military operation against foreign intervention was estimated in 1982 at $163 billion. For this reason, the Carter Doctrine was narrowly interpreted to mean defence against Soviet intervention and not intervention by one Gulf country against another (U.S. Congress, House Committee on Foreign Affairs, "U.S. Policy Toward the Persian Gulf," *Hearings Before the Subcommittee on Europe and the Middle East,* pp. 21, 28).

19. For a typical example of the criticism of the American-Saudi relationship during the Carter administration, see Sadiq Jalal al ʿAzm, *Siyast Carter wa Munazziru al-Haqba al-Saudiya* [Carter's Policy and Supporters of the Saudi Position] (Bayrut, 1977).

20. See Abdul Kasim Mansur [pen name of a former State Department official], "The American Threat to Saudi Arabia," *American Armed Forces Journal International,* September 1980, pp. 47–80; Christopher Van Hollen, "Don't Engulf the Gulf," *Foreign Affairs* (Summer 1981): 1064–78; Robert W. Tucker, "American Power and the Persian Gulf," *Commentary,* November 1980, pp. 25–41.

21. With regard to suggestions that some items of the Saudi request might be withheld or their use restricted, the Amir Sultan, the Saudi Minister of Defence, said, "Saudi Arabia cannot accept refusal of any of these items." "The Saudis," said Haig, "informed us that they could not accept some of the proposed restrictions on use of the AWACS" (Haig, *Caveat,* pp. 168, 188).

22. For the controversy over the arms sales to Saudi Arabia, see U.S. Congress, House Committee on Foreign Affairs, *Hearings and Markup Before the Committee on Foreign Affairs and the Subcommittee on International Security and Scientific Affairs and on Europe and the Middle East,* 97th Cong., 1st sess. (Washington, D.C., 1981).

23. See Haig, *Caveat,* pp. 189–90.

24. Ibid., p. 190. For an interpretation that this statement did not imply an American intention to intervene in Saudi internal affairs, see note 18 above.

25. For the sales of missiles, see U.S. Congress, House Committee on Foreign Affairs, *Hearings and Markup Before the Committee on Foreign Affairs and the*

*Subcommittee on Europe and the Middle East,* 99th Cong., 2nd sess., 22–23 April 1986 (Washington, D.C., 1986).

26. For a study of the structure and functions of the GCC, see ʿAbd-Allah Fahd al-Nafisi, *Majlis al-Taʾawin al-Khaliji* [The Gulf Cooperation Council] (London, 1982); Ursula Braun, *Der Kooperationsrat Arabischer Staaten am Golf: Eine Neue Kraft?* (Baden-Baden, 1986); Emile Nakhle, *The Arab Gulf Cooperative Council* (New York, 1986); John A. Sandwick, ed., *The Gulf Cooperation Council: Moderation and Stability in an Interdependent World* (Boulder, Colo., 1987).

27. For the steps taken to establish the United Arab Amirates, see Majid Khadduri, *Arab Personalities in Politics* (Washington, D.C., 1981), chap. 7.

28. The classic example of the Arab League's settlement of a dispute between two Gulf countries took place in 1961, when Iraq threatened Kuwayt by annexation on historical grounds (for a discussion of Iraq's claim to sovereignty over Kuwayt, see Majid Khadduri, *Republican Iraq* [London, 1969], pp. 166–73).

29. See John Christie, "History and Development of the Gulf Cooperation Council: A Brief Overview," in Sandwick, ed., *Gulf Cooperation Council,* pp. 18–19; John E. Peterson, "The GCC and Regional Security," in ibid., pp. 195–97.

30. See Fred H. Lawson, "Positive Sanctions and the Managing of International Sanctions," *International Journal* 40 (Autumn 1985): 628–54.

31. Kuwayt has increasingly become the area of subversive activities almost every year since the war began because of its financial and other assistance to Iraq as well as its proximity to the theater of operations. Between 1983 and 1985, the American and French embassies became targets of attacks, and American and foreign citizens were killed after planes in which they were traveling were hijacked. Iraq's attacks on refineries and the terminal at Kharj Island prompted Iran to escalate attacks on tankers carrying oil from Arab Gulf countries. For Iraq's attacks on Kharj Island, see Qasim Ahmad al-Uraybi, "Impact of the Iraq Blockade on the Island of Kharj," *Defence Magazine* 1 (1986): 181–96.

32. For the text of the communiqué, see *al-Watan* (Masqat) 2 and 7 November 1985; and *al-Sharq al-Awsat* (London) 7 November 1985.

33. Dr. Isam al-Chalabi, Iraq Minister of Oil, interview with author, Baghdad, April 21 and 27, 1987. Early in 1987, Dr. Chalabi announced that "Iraq has the capacity to export more than two million barrels a day for the first time since the war with Iran broke out in 1980." *Baghdad Observer,* 1 June 1987.

## XII. Conclusion

1. For the full text, see Appendix IV.
2. United Nations, Security Council, Official Records, S/19045, August 14, 1987.
3. United Nations, Security Council, Official Records, S/19883, July 24, 1987.
4. Ibid.
5. United Nations, Security Council, Official Records, S/19031, August 11, 1987.
6. Ibid.

7. United Nations Records S/19045; United Nations, Security Council, Official Records, S/19049, August 17, 1987.

8. United Nations, General Assembly, Official Records, A/42/PV.6, September 22, 1987.

# APPENDIX I

# Boundary Treaty Between the Kingdom of Iraq and the Empire of Iran

Signed at Tehran, 4 July 1937

His Majesty the King of Iraq, of the one part,
    and
His Imperial Majesty the Shahinshah of Iran, of the other part,
    Sincerely desirous of strengthening the bonds of brotherly friendship and good understanding between the two States, and of settling definitively the question of the frontier between their two States, have decided to conclude the present Treaty and have to that end appointed as their Plenipotentiaries:
    His Majesty the King of Iraq;
    His Excellency Dr Naji Al-Asil, Minister for Foreign Affairs;
    His Imperial Majesty the Shahinshah of Iran;
    His Excellency Monsieur Enayatollah Samiy, Minister for Foreign Affairs;
    Who, having communicated their full powers, found in good and due form, have agreed as follows.

### Article 1

The High Contracting Parties are agreed that, subject to the amendment for which Article 2 of the Present Treaty provides, the following documents shall be deemed valid and binding, that is to say:

(a) The Turco-Persian Delimitation Protocol signed at Constantinople on 4 November 1913;
(b) The Minutes of the meetings of the 1914 Frontier Delimitation Commission.

    In virtue of the present Article, the frontier between the two States shall be as defined and traced by the aforesaid Commission, save insofar as otherwise provided in Article 2 hereinafter following.

## Article 2

At the extreme point of the island of Shuteit (being approximately latitude 30° 17′ 25″ North, longitude 48° 19′ 28″ East), the frontier shall run perpendicularly from low-water mark to the *thalweg* of the Shatt al-Arab, and shall follow the same as far as a point opposite the present Jetty No. 1 at Abadan (being approximately latitude 30° 20′ 8.4″ North, longitude 48° 16′ 13″ East). From this point, it shall return to low-water mark, and follow the frontier line indicated in the 1914 Minutes.

## Article 3

Upon the signature of the present Treaty, the High Contracting Parties shall appoint forthwith a commission to erect frontier marks at the points determined by the commission to which Article 1, paragraph (b), of the present Treaty relates, and to erect such further marks as it shall deem desirable.

The composition of the commission and its programme of work shall be determined by special arrangement between the two High Contracting Parties.

## Article 4

The provisions hereinafter following shall apply to the Shatt al-Arab from the point at which the land frontier of the two States enters the said river to the high seas:

(a) The Shatt al-Arab shall remain open on equal terms to the trading vessels of all countries. All dues levied shall be in the nature of payments for services rendered and shall be devoted exclusively to meeting in equitable manner the cost of upkeep, maintenance of navigability or improvement of the navigable channel and the approach to the Shatt al-Arab from the sea, or to expenditure incurred in the interests of navigation. The said dues shall be calculated on the basis of the official tonnage of vessels or their displacement or both.

(b) The Shatt al-Arab shall remain open for the passage of vessels of war and other vessels of the two High Contracting Parties not engaged in trade.

(c) The circumstance that the frontier in the Shatt al-Arab sometimes follows the low-water mark and sometimes the *thalweg* or *medium filum aquae* shall not in any way affect the two High Contracting Parties' right of use along the whole length of the river.

## Article 5

The two High Contracting Parties, having a common interest in the navigation of the Shatt al-Arab as defined in Article 4 of the present Treaty, undertake to conclude a Convention for the maintenance and improvement of the navigable channel, and for dredging, pilotage, collection of dues, health measures, measures for preventing smuggling, and all other questions concerning navigation in the Shatt al-Arab, as defined in Article 4 of the present Treaty.

## Article 6

The present Treaty shall be ratified and the instruments of ratification shall be exchanged at Baghdad as soon as possible. It shall come into force as from the date of such exchange.

In faith whereof the Plenipotentiaries of the two High Contracting Parties have signed the present Treaty.

Done at Tehran, in the Arabic, Persian and French languages; in case of disagreement, the French text shall prevail.

This fourth day of July, one thousand nine hundred and thirty-seven.

*(signed)* Naji al-Asil
Samiy

# PROTOCOL

At the moment of signing the Frontier Treaty between Iraq and Iran, the two High Contracting Parties are agreed as follows:

## I

The geographical co-ordinates designated approximately in Article 2 of the aforesaid Treaty shall be definitively determined by a commission of experts consisting of an equal number of members appointed by each of the High Contracting Parties.

The geographical co-ordinates thus definitively determined within the limits fixed in the aforesaid Article shall be recorded in Minutes, the which, after signature by the members of the said commission, shall form an integral part of the Frontier Treaty.

## II

The High Contracting Parties undertake to conclude the Convention to which Article 5 of the Treaty relates within one year from the entry into force of the Treaty.

In the event of the said Convention not being concluded within the year despite their utmost efforts, the said time limit may be extended by the High Contracting Parties by common accord.

The Imperial Government of Iran agrees that, during the period of one year to which the first paragraph of the present Article relates or the extension (if any) of such period, the Royal Government of Iraq shall be responsible as at present for all questions to be settled under the said Convention. The Royal Government of Iraq shall notify the Imperial Government of Iran every six months as to the works executed, dues collected, expenditure incurred or any other measures undertaken.

## III

Permission granted by either of the High Contracting Parties to a vessel of war or other public-service vessel not engaged in trade, belonging to a third State, to enter its own harbours on the Shatt al-Arab shall be deemed to have been granted by the other High Contracting Party in such sort that the vessels in question shall be entitled to use the waters of the latter for the purpose of navigating the Shatt al-Arab.

The High Contracting Party gaining such permission shall immediately notify the other High Contracting Party accordingly.

## IV

It is clearly understood, without prejudice to the rights of Iran in respect of the Shatt al-Arab, that nothing in this Treaty shall affect the rights of Iraq and the contractual obligations of the same *vis-à-vis* the British Government in respect of the Shatt al-Arab under Article 4 of the Treaty of 30 June 1930 and paragraph 7 of the annex thereto signed on the same date.

## V

The present Protocol shall be ratified at the same time as the Frontier Treaty, of which it shall form an annex and integral part. It shall come into force at the same time as the Treaty.

The present Protocol is drawn up in Arabic, Persian and French; in case of difference, the French text shall prevail.

Done at Tehran in duplicate, the fourth day of July, one thousand nine hundred and thirty seven.

*(signed)* Naji al-Asil
Samiy

# APPENDIX II

# The Algiers Accord
# 6 March 1975

During the convocation of the OPEC summit conference in the Algerian capital and upon the initiative of President Houari Boumedienne, the Shah of Iran and Saddam Hussein (Vice-Chairman of the Revolution Command Council) met twice and conducted lengthy talks on the relations between Iraq and Iran. These talks, attended by President Houari Boumedienne, were characterized by complete frankness and a sincere will from both parties to reach a final and permanent solution of all problems existing between their two countries in accordance with the principles of territorial integrity, border inviolability and noninterference in internal affairs.

The two High Contracting Parties have decided to:

*First*: Carry out a final delineation of their land boundaries in accordance with the Constantinople Protocol of 1913 and the Proceedings of the Border Delimitation Commission of 1914.

*Second*: Demarcate their river boundaries according to the *thalweg* line.

*Third*: Accordingly, the two parties shall restore security and mutual confidence along their joint borders. They shall also commit themselves to carry out a strict and effective observation of their joint borders so as to put a final end to all infiltrations of a subversive nature wherever they may come from.

*Fourth*: The two parties have also agreed to consider the aforesaid arrangements as inseparable elements of a comprehensive solution. Consequently, any infringement of one of its components shall naturally contradict the spirit of the Algiers Accord. The two parties shall remain in constant contact with President Houari Boumedienne who shall provide, when necessary, Algeria's brotherly assistance whenever needed in order to apply these resolutions.

The two parties have decided to restore the traditional ties of good neighbourliness and friendship, in particular by eliminating all negative factors in their relations and through constant exchange of views on issues of mutual interest and promotion of mutual co-operation.

The two parties officially declare that the region ought to be secure from any foreign interference.

The Foreign Ministers of Iraq and Iran shall meet in the presence of Algeria's Foreign Minister on 15 March 1975 in Tehran in order to make working arrangements for the Iraqi-Iranian joint commission which was set up to apply the resolutions taken by mutual agreement as specified above. And in accordance with the desire of the two parties, Algeria shall be invited to the meetings of the Iraqi-Iranian joint commission. The commission shall determine its agenda and working procedures and hold meetings if necessary. The meetings shall be alternately held in Baghdad and Tehran.

His Majesty the Shah of Iran accepted with pleasure the invitation extended to him by His Excellency President Ahmad Hasan al-Bakr to pay a state visit to Iraq. The date of the visit shall be fixed by mutual agreement.

On the other hand, Saddam Hussein agreed to visit Iran officially at a date to be fixed by the two parties.

H.M. the Shah of Iran and Saddam Hussein expressed their deep gratitude to President Houari Boumedienne, who, motivated by brotherly sentiments and a spirit of disinterestedness, worked for the establishment of a direct contact between the leaders of the two countries and consequently contributed to reviving a new era in the Iraqi-Iranian relations with a view to achieving the higher interest of the future of the region in question.

# APPENDIX III

# Treaty Concerning the Frontier and Neighbourly Relations Between Iran and Iraq

His Imperial Majesty the Shahinshah of Iran,

His Excellency the President of the Republic of Iraq,

Considering the sincere desire of the two Parties as expressed in the Algiers Agreement of 6 March 1975, to achieve a final and lasting solution to all the problems pending between the two countries,

Considering that the two Parties have carried out the definitive redemarcation of their land frontier on the basis of the Constantinople Protocol of 1913 and the minutes of the meetings of the Frontier Delimitation Commission of 1914 and have delimited their river frontier along the thalweg,

Considering their desire to restore security and mutual trust throughout the length of their common frontier,

Considering the ties of geographical proximity, history, religion, culture and civilization which bind the peoples of Iran and Iraq,

Desirous of strengthening their bonds of friendship and neighbourliness, expanding their economic and cultural relations and promoting exchange and human relations between their peoples on the basis of the principles of territorial integrity, the inviolability of frontiers and non-interference in internal affairs,

Resolved to work towards the introduction of a new era in friendly relations between Iran and Iraq based on full respect for the national independence and sovereign equality of States,

Convinced that they are helping thereby to implement the principles and achieve the purposes and objectives of the Charter of the United Nations,

Have decided to conclude this Treaty and have appointed as their plenipotentiaries:

**His Imperial Majesty the Shahinshah of Iran:**
His Excellency Abbas Ali Khalatbary, Minister of Foreign Affairs of Iran.

**His Excellency the President of the Republic of Iraq:**
His Excellency Saadoun Hamadi, Minister for Foreign Affairs of Iraq.

Who, having exchanged their full powers, found to be in good and due form, have agreed as follows:

## Article 1

The High Contracting Parties confirm that the State land frontier between Iraq and Iran shall be that which has been redemarcated on the basis of and in accordance with the provisions of the Protocol concerning the redemarcation of the land friend, and the annexes thereto, attached to this Treaty.

## Article 2

The High Contracting Parties confirm that the State frontier in the Shatt Al Arab shall be that which has been delimited on the basis of and in accordance with the provisions of the Protocol concerning the delimitation of the river frontier, and the annexes thereto, attached to this Treaty.

## Article 3

The High Contracting Parties undertake to exercise strict and effective permanent control over the frontier in order to put an end to any infiltration of a subversive nature from any source, on the basis of and in accordance with the provisions of the Protocol concerning frontier security, and the annex thereto, attached to this Treaty.

## Article 4

The High Contracting Parties confirm that the provisions of the three Protocols, and the annexes thereto, referred to in articles 1, 2, and 3 above and attached to this Treaty as an integral part thereof shall be final and permanent. They shall not be infringed under any circumstances and shall constitute the indivisible elements of an over-all settlement. Accordingly, a breach of any of the components of this over-all settlement shall clearly be incompatible with the spirit of the Algiers Agreement.

## Article 5

In keeping with the inviolability of the frontiers of the two States and strict respect for their territorial integrity, the High contracting Parties confirm that the course of their land and river frontiers shall be inviolable, permanent and final.

## Article 6

1. In the event of a dispute regarding the interpretation or implementation of this Treaty, the three Protocols or the annexes thereto, any solution to such a dispute shall strictly respect the course of the Iraqi-Iranian frontier referred to in articles 1

and 2 above, and shall take into account the need to maintain security on the Iraqi-Iranian frontier in accordance with article 3 above.

2. Such disputes shall be resolved in the first instance by the High Contracting Parties, by means of direct bilateral negotiations to be held within two months after the date on which one of the Parties so requested.

3. If no agreement is reached, the High Contracting Parties shall have recourse, within a three-month period, to the good offices of a friendly third State.

4. Should one of the two Parties refuse to have recourse to good offices or should the good offices procedure fail, the dispute shall be settled by arbitration within a period of not more than one month after the date of such refusal or failure.

5. Should the High Contracting Parties disagree as to the arbitration procedure, one of the High Contracting Parties may have recourse, within 15 days after such disagreement was recorded, to a court of arbitration.

With a view to establish such a court of arbitration each of the High Contracting Parties shall, in respect of each dispute to be resolved, appoint one of its nationals as arbitrators and the two arbitrators shall choose an umpire. Should the High Contracting Parties fail to appoint their arbitrators within one month after the date on which one of the Parties received a request for arbitration from the other Party, or should the arbitrators fail to reach agreement on the choice of the umpire before that time-limit expires, the High Contracting Party which requested arbitration shall be entitled to request the President of the International Court of Justice to appoint the arbitrators or the umpire, in accordance with the procedures of the Permanent Court of Arbitration.

6. The decision of the court of arbitration shall be binding on and enforceable by the High Contracting Parties.

7. The High Contracting Parties shall each defray half the costs of arbitration.

## Article 7

This Treaty, the three Protocols and the annexes thereto shall be registered in accordance with Article 102 of the Charter of the United Nations.

## Article 8

This Treaty, the three Protocols and the annexes thereto shall be ratified by each of the High Contracting Parties in accordance with its domestic law.

This Treaty, the three Protocols and the annexes thereto shall enter into force on the date of the exchange of the instruments of ratification in Teheran.

IN WITNESS WHEREOF the Plenipotentiaries of the High Contracting Parties have signed this Treaty, the three Protocols and the annexes thereto.

DONE at Baghdad, on 13 June 1975.

*(Signed)*                                                    *(Signed)*

Abbas Ali Khalatbary                                Saadoun Hamadi

Minister for Foreign Affairs of Iran         Minister for Foreign Affairs of Iraq

This Treaty, the three Protocols and the annexes thereto were signed in the presence of His Excellency Abdel-Aziz Bouteflika, Member of the Council of the Revolution and Minister for Foreign Affairs of Algeria

*(Signed)*

# PROTOCOL CONCERNING THE DELIMITATION TO THE RIVER FRONTIER BETWEEN IRAN AND IRAQ

Pursuant to the decisions taken in the Algiers communiqué of 6 March 1975,
The two Contracting Parties have agreed as follows:

## Article 1

The two Contracting Parties hereby declare and recognize that the State river frontier between Iran and Iraq in the Shatt Al Arab has been delimited along the thalweg by the Mixed Iraqi-Iranian-Algerian Committee on the basis of the following:

1. The Teheran Protocol of 17 March 1975;
2. The record of the Meeting of Ministers for Foreign Affairs, signed at Baghdad on 20 April 1975, approving, *inter alia,* the record of the Committee to Delimit the River Frontier, signed on 16 April 1975 on board the Iraqi ship *El Thawra* in the Shatt Al Arab;
3. Common hydrographic charts, which have been verified on the spot and corrected and on which the geographical co-ordinates of the 1975 frontier crossing points have been indicated; these charts have been signed by the hydrographic experts of the Mixed Technical Commission and countersigned by the heads of the Iran, Iraq and Algerian delegations to the Committee. The said charts, listed hereinafter, are annexed to this Protocol and form an integral part thereof:

Chart No. 1: Entrance to the Shatt Al Arab, No. 3842, published by the British Admiralty;
Chart No. 2: Inner Bar to Kabda Point, No. 3843, published by the British Admiralty;
Chart No. 3: Kabda Point to Abadan, No. 3844, published by the British Admiralty;
Chart No. 4: Abadan to Jazirat Ummat Tuwaylah, No. 3845, published by the British Admiralty.

## Article 2

1. The frontier line in the Shatt Al Arab shall follow the thalweg, i.e., the median line of the main navigable channel at the lowest navigable level, starting from the

point at which the land frontier between Iran and Iraq enters the Shatt Al Arab and continuing to the sea.

2. The frontier line, as defined in paragraph 1 above, shall vary with changes brought about by natural causes in the main navigable channel. The frontier line shall not be affected by other changes unless the two Contracting Parties conclude a special agreement to that effect.

3. The occurrence of any of the changes referred to in paragraph 2 above shall be attested jointly by the competent technical authorities of the two Contracting Parties.

4. Any change in the bed of the Shatt Al Arab brought about by natural causes which would involve a change in the national character of the two States' respective territory or of landed property, constructions, or technical or other installations shall not change the course of the frontier line, which shall continue to follow the thalweg in accordance with the provisions of paragraph 1 above.

5. Unless an agreement is reached between the two Contracting Parties concerning the transfer of the frontier line to the new bed, the water shall be re-directed at the joint expense of both Parties to the bed existing in 1975—as marked on the four common charts listed in article 1, paragraph 3, above—should one of the Parties so request within two years after the date on which the occurrence of the change was attested by either of the two Parties. Until such time, both Parties shall retain their previous rights of navigation and of use over the water of the new bed.

## Article 3

1. The river frontier between Iran and Iraq in the Shatt Al Arab, as defined in article 2 above, is represented by the relevant line drawn on the common charts referred to in article 1, paragraph 3, above.

2. The two Contracting Parties have agreed to consider that the river frontier shall end at the straight line connecting the two banks of the Shatt Al Arab, at its mouth, at the astronomical lowest low-water mark. This straight line has been indicated on the common hydrographic charts referred to in article 1, paragraph 3, above.

## Article 4

The frontier line as defined in articles 1, 2 and 3 of this Protocol shall also divide vertically the air space and the subsoil.

## Article 5

With a view to eliminating any source of controversy, the two Contracting Parties shall establish a Mixed Iraqi-Iranian Commission to settle, within two months, any questions concerning the status of landed property, constructions, or technical or other installations, the national character of which may be affected by the delimitation of the Iranian-Iraqi river frontier, either through repurchase or compensation or any other suitable arrangement.

## Article 6

Since the task of surveying the Shatt Al Arab has been completed and the common hydrographic chart referred to in article 1, paragraph 3, above has been drawn up, the two Contracting Parties have agreed that a new survey of the Shatt Al Arab shall be carried out jointly, once every 10 years, with effect from the date of signature of this Protocol. However, each of the two Parties shall have the right to request new surveys, to be carried out jointly, before the expiry of the 10-year period.

The two Contracting Parties shall each defray half the cost of such surveys.

## Article 7

1. Merchant vessels, State vessels and warships of the two Contracting Parties shall enjoy freedom of navigation in the Shatt Al Arab and in any part of the navigable channels in the territorial sea which lead to the mouth of the Shatt Al Arab, irrespective of the line delimiting the territorial sea of each of the two countries.

2. Vessels of third countries used for purposes of trade shall enjoy freedom of navigation, on an equal and non-discriminatory basis, in the Shatt Al Arab and in any part of the navigable channels in the territorial sea which lead to the mouth of the Shatt Al Arab, irrespective of the line delimiting the territorial sea of each of the two countries.

3. Either of the two Contracting Parties may authorize frontier warships visiting its ports to enter the Shatt Al Arab, provided such vessels do not belong to a country in a state of belligerency, armed conflict or war with either of the two Contracting Parties and provided the other Party is so notified no less than 72 hours in advance.

4. The two Contracting Parties shall in every case refrain from authorizing the entry to the Shatt Al Arab of merchant vessels belonging to a country in a state of belligerency, armed conflict or war with either of the two Parties.

## Article 8

1. Rules governing navigation in the Shatt Al Arab shall be drawn up by a mixed Iranian-Iraqi Commission, in accordance with the principle of equal rights of navigation for both States.

2. The two Contracting Parties shall establish a Commission to draw up rules governing the prevention and control of pollution in the Shatt Al Arab.

3. The two Contracting Parties undertake to conclude subsequent agreements on the questions referred to in paragraphs 1 and 2 of this article.

## Article 9

The two Contracting Parties recognize that the Shatt Al Arab is primarily an international waterway, and undertake to refrain from any operation that might hinder navigation in the Shatt Al Arab or in any part of those navigable channels in the territorial sea of either of the two countries that lead to the mouth of the Shatt Al Arab.

DONE at Baghdad, on 13 June 1975.

*(Signed)*                                                                              *(Signed)*

Abbas Ali Khalatbary                                                     Saadoun Hamadi
Minister for Foreign Affairs of Iran          Minister for Foreign Affairs of Iraq

Signed in the presence of His Excellency Abdel-Aziz Bouteflika, Member of the Council Revolution and Minister for Foreign Affairs of Algeria

*(Signed)*

# APPENDIX IV

# United Nations Resolutions on the Iraq-Iran War The Situation Between Iraq and Iran

## Decisions

On 23 September 1980, the president of the Council issued the following statement:

Members of the Security Council have today exchanged views in informal consultations on the extremely serious situation prevailing between Iran and Iraq. They have taken note of the sharp deterioration in relations and of the escalation in armed activity leading to loss of life and heavy material damage.

Members of the Council are deeply concerned that this conflict can prove increasingly serious and could pose a grave threat to international peace and security.

Members of the Council welcome and fully support the appeal of the Secretary-General, addressed to both parties on 22 September 1980, as well as the offer that he has made of his good offices to resolve the present conflict.

The members of the Council have asked me to appeal, on their behalf to the Governments of Iran and Iraq, as a first step towards a solution of the conflict, to desist from all armed activity and all acts that may worsen the present dangerous situation and to settle their dispute by peaceful means.

At its 2247th meeting, on 26 September 1980, the Council decided to invite the representative of Iraq to participate, without vote, in the discussion of the item entitled "the situation between Iran and Iraq."

At its 2248th meeting on 28 September 1980, the Council decided to invite the representative of Japan to participate, without vote, in the discussion of the question.

## (1) Resolution 479 of 28 September 1980

The Security Council

Having begun consideration of the item entitled "The situation between Iran and Iraq,"

Mindful that all Member States have undertaken, under the Charter of the United Nations, the obligation to settle their international disputes by peaceful means and in such a manner that international peace and security and justice are not endangered,

Mindful as well that all Member States are obliged to refrain their international relations from the threat of or use of force against the terrritorial integrity or political independence of any State,

Recalling that under Article 24 of the Charter the Security Council has primary responsibility for the maintenance of international peace and security,

Deeply concerned about the developing situation between Iran and Iraq,

1. Calls upon Iran and Iraq to refrain immediately from any further use of force and to settle their dispute by peaceful means in conformity with principles of justice and international law;

2. Urges them to accept any appropriate offer of mediation or conciliation or to resort to regional agencies or arrangements or other peaceful means of their own choice that would facilitate the fulfillment of their obligations under the Charter of the United Nations;

3. Calls upon all other States to exercise the utmost restraint and to refrain from any act which may lead to a further escalation and widening of the conflict;

4. Supports the efforts of the Secretary-General and the offer of his good offices for the resolution of this situation;

5. Requests the Secretary-General to report to the Security Council within forty-eight hours.

Adopted unanimously at the 2248th meeting

## (2) Resolution 514 (1982) Adopted by the Security Council at its 2383rd meeting on 12 July 1982

The Security Council,

Having considered again the question entitled "The situation between Iran and Iraq,"

Deeply concerned about the prolongation of the conflict between the two countries, resulting in heavy losses of human lives and considerable material damage, and endangering peace and security,

Recalling the provisions of Article 2 of the Charter of the United Nations, and that the establishment of peace and security in the region requires strict adherence to these provisions,

Recalling that by virtue of Article 24 of the Charter the Security Council has the primary responsibility for maintenance of international peace and security,

Recalling its resolution 479 (1980), adopted unanimously on 28 September 1980, as well as the statement of its President of 5 November 1980 (S/14244),

Taking note of the efforts of mediation pursued notably by the Secretary-General of the United Nations and his representative, as well as by the Movement of Non-Aligned Countries and the Organization of the Islamic Conference,

1. Calls for a cease-fire and an immediate end to all military operations;

2. Calls further for a withdrawal of forces to internationally recognized boundaries;

3. Decides to dispatch a team of United Nations observers to verify, confirm and supervise the cease-fire and withdrawal, and requests the Secretary-General to submit to the Council a report on the arrangements required for that purpose;

4. Urges that the mediation efforts be continued in a coordinated manner through the Secretary-General with a view to achieving a comprehensive, just and honorable settlement acceptable to both sides of all the outstanding issues, on the basis of the principles of the Charter of the United Nations, including respect for sovereignty, independence, territorial integrity and non-interference in the internal affairs of States;

5. Requests all other States to abstain from all actions which could contribute to the continuation of the conflict and to facilitate the implementation of the present resolution;

6. Requests the Secretary-General to report to the Security Council within three months on the implementation of this resolution.

### (3) Consequences of the Prolongation of the Armed Conflict between Iran and Iraq

The General Assembly,

Having considered the item entitled "Consequences of the prolongation of the armed conflict between Iran and Iraq,"

Noting the Preamble of the Charter of the United Nations, in which all States expressed their determination to live together in peace with one another as good neighbours,

Reaffirming the principles that no State should acquire or occupy territories by the use of force, that whatever territories had been acquired in this way should be returned, that no act of aggression should be committed against any State, that the territorial integrity and the sovereignty of all States should be respected, that no State should try to interfere or intervene in the internal affairs of other States and that all differences or claims which may exist between States should be settled by peaceful means in order that peaceful relations should prevail among Member States,

Recalling resolutions 479 (1980) of 28 September 1980, 514 (1982) of 12 July 1982 and 522 (1982) of 4 October 1982 on the question entitled "The situation between Iran and Iraq," unanimously adopted by the Security Council,

Further recalling the statements made by the President of the Security Council on 5 November 1980 and 15 July 1982,

Taking note of the report of the Secretary-General of 7 October 1982,

Considering that the Security Council has already called for an immediate cease-fire and an end to all military operations,

Considering further that the prolongation of the conflict constitutes a violation of the obligations of Member States under the Charter,

1. Considers that the conflict between Iran and Iraq and its prolongation and recent escalation, resulting heavy losses in human lives and considerable material dam-

age in a politically and economically strategic region, endanger international peace and security;

2. Affirms the necessity of achieving an immediate cease-fire and withdrawal of forces to internationally recognized boundaries as a preliminary step towards the settlement of the dispute by peaceful means in conformity with the principles of justice and international law;

3. Calls upon all other States to abstain from all actions which could contribute to the continuation of the conflict and to facilitate the implementation of the present resolution;

4. Requests the Secretary-General to continue his efforts, in consultation with the parties concerned, with a view to achieving a peaceful settlement;

5. Further requests the Secretary-General to keep Member States informed on the implementation of the present resolution.

41st plenary meeting
22 October 1982

## (4) Co-operation between the United Nations and the Organization of the Islamic Conference

The General Assembly

Having considered the report of the Secretary-General on co-operation between the United Nations and the Organization of the Islamic Conference,

Recalling its resolution 3369 (XXX) of 10 October 1975, by which it granted observer status to the Organization of the Islamic Conference,

Recalling its resolutions 35/36 of 14 November 1980 and 36/23 of 9 November 1981,

Noting with satisfaction the continued development of co-operation between the United Nations and the Organization of the Islamic Conference,

Noting the strengthening of co-operation between the specialized agencies and other organizations of the United Nations system and the Organization of the Islamic Conference,

Taking into account the desire of both organizations to co-operate more closely in their common search for solutions to global problems, such as questions relating to international peace and security, disarmament, self-determination, decolonization, fundamental human rights and the establishment of a new international economic order,

Noting also the signing of co-operation agreements between a number of specialized agencies and the Organization of the Islamic Conference,

Convinced of the need to strengthen further the cooperation between the United Nations and the Organization of the Islamic Conference,

Noting further the proposals of the Secretary-General,

1. Takes note with satisfaction of the report of the Secretary-General and endorses the proposals contained therein;

2. Requests the United Nations and the Organization of the Islamic Conference to

intensify co-operation in their common search for solutions to global problems, such as questions relating to international peace and security, disarmament, self-determination, decolonization, fundamental human rights and the establishment of a new international economic order;

3. Requests the Secretary-General to prepare guidelines based on resolutions of the General Assembly for promoting co-operation with the Organization of the Islamic Conference;

4. Invites the Secretary-General, in consultation with the Secretary-General of the Organization of the Islamic Conference, to organize an annual meeting, beginning in 1983, between the secretariat of the Organization of the Islamic Conference and the secretariats of the United Nations and other organizations concerned within the United Nations system to examine the stage reached in the development of co-operation and to put forward proposals for promoting co-operation with the Organization of the Islamic Conference;

5. Encourages the specialized agencies and other organizations concerned within the United Nations system to continue to expand their co-operation with the Organization of the Islamic Conference, *inter alia* by negotiating co-operation agreements;

6. Requests the Secretary-General to continue to take steps to strengthen the co-ordination of the activities of the United Nations system in this field with a view to intensifying co-operation between the United Nations and the United Nations system and the Organization of the Islamic Conference;

7. Calls upon the Secretary-General to report to the General Assembly at its thirty-eighth session on the state of co-operation between the United Nations and the Organization of the Islamic Conference;

8. Decides to include in the provisional agenda of its thirty-eighth session the item entitled "Co-operation between the United Nations and the Organization of the Islamic Conference."

41st plenary meeting
22 October 1982

## (5) Resolution 522 (1982) Adopted by the Security Council at its 2399th meeting on 4 October 1982

The Security Council,

Having considered again the question entitled "The situation between Iran and Iraq,"

Deploring the prolongation and the escalation of the conflict between the two countries, resulting in heavy losses of human lives and considerable material damage, and endangering peace and security,

Reaffirming that the restoration of peace and security in the region requires all Member States strictly to comply with their obligations under the Charter of the United Nations,

Recalling its resolution 479 (1980), adopted unanimously on 28 September 1980, as well as the statement of the President of the Council of 5 November 1980 (S/14244),

Further recalling its resolution 514 (1982), adopted unanimously on 12 July 1982 and the statement of the President of the Council of 15 July 1982 (S/15295),

Taking note of the report of the Secretary-General (S/15293) of 15 July 1982,

1. Urgently calls again for an immediate cease-fire and an end to all military operations;

2. Reaffirms its call for a withdrawal of forces to internationally recognized boundaries;

3. Welcomes the fact that one of the parties has already expressed its readiness to co-operate in the implementation of resolution 514 (1982) and calls upon the other to do likewise;

4. Affirms the necessity of implementing without further delay its decision to dispatch United Nations observers to verify, confirm and supervise the cease-fire and withdrawal;

5. Reaffirms the urgency of the continuation of the current mediation efforts;

6. Reaffirms its request to all other States to abstain from all actions which could contribute to the continuation of the conflict and to facilitate the implementation of the present resolution;

7. Further requests the Secretary-General to report to the Council on the implementation of this resolution within 72 hours.

## (6) Resolution 540 (1983) Adopted by the Security Council at its 2493rd meeting on 31 October 1983

The Security Council,

Having considered again the question "The situation between Iran and Iraq,"

Recalling its relevant resolutions and statements which, *inter alia,* call for a comprehensive cease-fire and an end to all military operations between the parties,

Recalling the report of the Secretary-General of 20 June 1983 (S/15834) on the mission appointed by him to inspect civilian areas in Iran and Iraq which have been subject to military attacks, and expressing its appreciation to the Secretary-General for presenting a factual, balanced and objective account,

Also noting with appreciation and encouragement the assistance and co-operation given to the Secretary-General's mission by the Governments of Iran and Iraq,

Deploring once again the conflict between the two countries, resulting in heavy losses of civilian lives and extensive damage caused to cities, property and economic infrastructures,

Affirming the desirability of an objective examination of the causes of the war,

1. Requests the Secretary-General to continue his mediation efforts with the parties concerned, with a view to achieving a comprehensive, just and honorable settlement acceptable to both sides;

2. Condemns all violations of international humanitarian law, in particular, the provisions of the Geneva Conventions of 1949 in all their aspects, and calls for the

immediate cessation of all military operations against civilian targets, including city and residential areas;

3. Affirms the right of free navigation and commerce in international waters, calls on all States to respect this right and also calls upon the belligerents to cease immediately all hostilities in the region of the Gulf, including all sea-lanes, navigable waterways, harbour works, terminals, offshore installations and all ports with direct or indirect access to the sea, and to respect the integrity of the other littoral States;

4. Requests the Secretary-General to consult with the parties concerning ways to sustain and verify the cessation of hostilities, including the possible dispatch of United Nations observers, and to submit a report to the Council on the results of these consultations;

5. Calls upon both parties to refrain from any action that may endanger peace and security as well as marine life in the region of the Gulf;

6. Calls once more upon all other States to exercise the utmost restraint and to refrain from any act which may lead to a further escalation and widening of the conflict and, thus, to facilitate the implementation of the present resolution;

7. Requests the Secretary-General to consult with the parties regarding immediate and effective implementation of this resolution.


## (7) Resolution 552 (1984) Adopted by the Security Council at its 2546th meeting on 1 June 1984

The Security Council,

Having considered the letter dated 21 May 1984 from the representatives of Bahrain, Kuwait, Oman, Qatar, Saudi Arabia and the United Arab Emirates (S/16574) complaining against Iranian attacks on commercial ships en route to and from the ports of Kuwait and Saudi Arabia,

Noting that Member States pledged to live together in peace with one another as good neighbors in accordance with the United Nations Charter,

Reaffirming the obligations of Member States to the principles and purposes of the United Nations Charter,

Reaffirming also that all Member States are obliged to refrain in their international relations from the threat or use of force against the territorial integrity or political independence of any State,

Taking into consideration the importance of the Gulf region to international peace and security and its vital role to the stability of world economy,

Deeply concerned over the recent attacks on commercial ships en route to and from the ports of Kuwait and Saudi Arabia,

Convinced that these attacks constitute a threat to the safety and stability of the area and have serious implications for international peace and security,

1. Calls upon all States to respect, in accordance with international law, the right of free navigation;

2. Reaffirms the right of free navigation in international withdrawals and sea lanes

for shipping en route to and from all ports and installations of the littoral States that are not parties to the hostilities;

3. Calls upon all States to respect the territorial integrity of the States that are not parties to the hostilities and to exercise the utmost restraint and to refrain from any act which may lead to a further escalation and widening of the conflict;

4. Condemns these recent attacks on commercial ships en route to and from the ports of Kuwait and Saudi Arabia;

5. Demands that such attacks should cease forthwith and that there should be no interference with ships en route to and from States that are not parties to the hostilities;

6. Decides, in the event of non-compliance with the present resolution, to meet again to consider effective measures that are commensurate with the gravity of the situation in order to ensure the freedom of navigation in the area;

7. Requests the Secretary-General to report on the progress of the implementation of the present resolution;

8. Decides to remain seized of the matter.

## Resolution 598 (July 20, 1987)

The Security Council,

Reaffirming its resolution 582 (1986)

Deeply concerned that, despite its calls for a cease-fire, the conflict between Iran and Iraq continues unabated, with further heavy loss of human life and material destruction,

Deploring the initiation and continuation of the conflict,

Deploring also the bombing of purely civilian population centers, attacks on neutral shipping or civilian aircraft, the violation of international humanitarian law and other laws of armed conflict, and, in particular, the use of chemical weapons contrary to obligations under the 1925 Geneva Protocol,

Deeply concerned that further escalation and widening of the conflict may take place,

Determined to bring to an end all military actions between Iran and Iraq,

Convinced that a comprehensive, just, honorable and durable settlement should be achieved between Iran and Iraq,

Recalling the provisions of the Charter of the United Nations and in particular the obligation of all member states to settle their international disputes by peaceful means in such a manner that international peace and security and justice are not endangered,

Determining that there exists a breach of the peace as regards the conflict between Iran and Iraq,

Acting under Articles 39 and 40 of the Charter of the United Nations,

1. Demands that, as a first step toward a negotiated settlement, Iran and Iraq observe an immediate cease-fire, discontinue all military actions on land, at sea and in the air, and withdraw all forces to the internationally recognized boundaries without delay;

2. Requests the Secretary-General to dispatch a team of United Nations observers to verify, confirm and supervise the cease-fire and withdrawal and further requests the Secretary-General to make the necessary arrangements in consultation with the parties and to submit a report thereon to the Security Council;

3. Urges that prisoners of war be released and repatriated without delay after the cessation of active hostilities in accordance with the Third Geneva Convention of 12 August 1949;

4. Calls upon Iran and Iraq to cooperate with the Secretary General in implementing this resolution and in mediation efforts to achieve a comprehensive, just and honorable settlement, acceptable to both sides, of all outstanding issues in accordance with the principles contained in the Charter of the United Nations;

5. Calls upon all other states to exercise the utmost restraint and to refrain from any act which may lead to further escalation and widening of the conflict and thus to facilitate the implementation of the present resolution;

6. Requests the Secretary General to explore, in consultation with Iran and Iraq, the question of entrusting an impartial body with inquiring into responsibility for the conflict and to report to the Security Council as soon as possible;

7. Recognizes the magnitude of the damage inflicted during the conflict and the need for reconstruction efforts with appropriate international assistance once the conflict is ended and in this regard requests the Secretary General to assign a team of experts to study the question of reconstruction and to report to the Security Council;

8. Further requests the Secretary General to examine in consultation with Iran and Iraq and with other states of the region measures to enhance the security and stability of the region;

9. Requests the Secretary General to keep the Security Council informed on the implementation of this resolution;

10. Decides to meet again as necessary to consider further steps to insure compliance with this resolution.

# APPENDIX V

# Replies of Iraq and Iran to Security Council Resolution 598 (1987)

## Iraq's Reply to Resolution 598

Letter dated 23 July 1987 from the Deputy Prime Minister and Minister for Foreign Affairs of Iraq addressed to the Secretary-General

I have the honour to refer to your letter dated 20 July 1987 and to inform you that the Iraqi Government has studied the text of Security Council resolution 598 (1987), adopted unanimously by the Council on 20 July 1987. The President of the Republic of Iraq has instructed me to transmit to you the position of the Iraqi Government, which is as follows:

1. The Iraqi Government welcomes the resolution and is ready to co-operate with you and with the Security Council so as to implement it in good faith with a view to finding a comprehensive, just, lasting and honourable settlement of the conflict with Iran.

2. On the basis of the contents of the resolution and its binding character under Chapter VII of the Charter, it is, of course, obvious that Iran's clear approval of the resolution, confirmed to you, and its clear readiness to fulfil its obligations thereunder, without any terms or conditions, in good faith and with serious intent, are essential for the fulfilment of the corresponding obligations which rest upon us. In that regard, I have set forth in paragraph 1 above our complete readiness to co-operate with you in fulfilling those obligations in good faith with a view to finding a comprehensive, just, lasting and honourable settlement of the conflict.

3. In welcoming the resolution, the Iraqi Government proceeds from the premise that the text thereof is an integral and indivisible whole in respect of the contents, the time-limits and the measures for the implementation of all its paragraphs, and in particular from the premise of immediate and mutual advantage from its implementation to all the parties concerned.

4. The Iraqi Government takes the expression "without delay," which appears in paragraph 1 of the resolution, to mean that the withdrawal shall be completed within a period not exceeding 10 days from the date of the general cease-fire. The determination of this period derives from the precedent of the withdrawal of Iraqi forces from Iranian territory, which was effected within 10 days, between 10 and 20 June 1982, even though that withdrawal was unilateral and took place in the absence of a cease-fire from the other side.

5. The Iraqi Government takes the expression "without delay," which appears in paragraph 3 of the resolution and concerns the release and repatriation of prisoners-of-war, to mean that the prisoners shall be released and repatriated within a period not exceeding eight weeks from the date of the cease-fire. Furthermore, the Iraqi Government understands that this operation shall, in accordance with the third Geneva Convention of 12 August 1949, be effected in co-operation with the International Committee of the Red Cross, and that this is a humanitarian and moral question having no connection in any way with the negotiations on other matters pending between the two countries.

6. The Iraqi Government is ready to co-operate with you sincerely in the mediation efforts entrusted to you in order to achieve a comprehensive, just and honourable settlement in accordance with the principles set forth in the Charter of the United Nations.

The Iraqi Government will submit its proposals and define its position on other pending questions when negotiations commence concerning the comprehensive settlement, as called for in the resolution of the Security Council.

7. Iraq understands that, as soon as the cease-fire begins, it will be able to utilize its ports, its coasts and its internal and territorial waters and also that it will be able to enjoy, on a footing of equality with Iran, freedom of navigation in the international waters of the Arabian Gulf.

8. With regard to the provisions of paragraph 6 of the resolution, the Iraqi Government wishes to emphasize that it is ready to engage in consultations with you concerning the inquiry into responsibility for the conflict and its protraction and concerning the body to which this task should be entrusted.

9. The Iraqi Government welcomes the contents of paragraph 8 concerning measures to enhance the security and stability of the region, and proposes that, in the stage following the establishment of peace between Iraq and Iran, you convene a meeting of the Ministers for Foreign Affairs of the States of the Arabian Gulf, under your auspices, to study ways and means of guaranteeing security, stability and the freedom of international navigation in the region of the Arabian Gulf, on the basis of full respect for the sovereignty of the States concerned, non-intervention in each other's internal affairs and observance of the provisions of international law.

10. Lastly, the Iraqi Government, in keeping with the nature of the resolution adopted by the Council and in the light of the strong desire of the international community to bring about peace as a matter of urgency, hopes that the period of time required for the submission of your report to the Security Council on the imple-

mentation of the resolution, in pursuance of paragraph 9, will be short so as to prevent any procrastination or delay from any quarter whatsoever.

*(Signed)* Tariq AZIZ
Deputy Prime Minister
Minister for Foreign Affairs
of the Republic of Iraq

## Iran's Reply to Resolution 598

### Detailed and official position of the Islamic Republic of Iran on the Security Council resolution 598 (1987)

1. Resolution 598 (1987) has been formulated and adopted by the United States with the explicit intention of intervention in the Persian Gulf and the region, mustering support for Iraq and its supporters in the war, and the diversion of public opinion from the home front. None of these objectives correspond to the legitimate objective of seeking a just solution to the conflict.

2. Resolution 598 (1987) has been formulated without seeking consultation from the Islamic Republic of Iran. As it reflects the Iraqi formulae for the resolution of the conflict, it cannot therefore be considered a balanced, impartial, comprehensive and practical resolution.

3. If the United States and the countries supporting Iraq harbour the illusion that they can terminate the war in favour of Iraq through an unjust and partial resolution, they are well advised to review the experience of the Council over the past several years following the liberation of Khorramshahr and the expulsion of the Iraq forces of occupation from most of the occupied territory inside the Islamic Republic of Iran. The Islamic Republic of Iran, as the victim of aggression, is the main party to determine how the war can be terminated, and no change can be effected in the course of the war as long as conditions of the Islamic Republic of Iran are not met.

4. The Security Council is obliged to explain why the Iran-Iraq war, exactly at the time when it is approaching its final stages, has turned into a breach of peace, thus necessitating recourse to Article 39 of the Charter. Ironically enough, the initiation of the war by Iraq on 22 September 1980 and occupation of a vast part of the territory of the Islamic Republic of Iran was a breach of world peace. The Security Council, however, chose to remain silent then. Little wonder, then, that the United States, in an outright show of support for Iraq even forced the Security Council to oppose the amendment of some permanent and non-permanent members of the Council as to considering the war from the very outset as a breach of peace.

5. The first Article of the United Nations Charter accords "suppression of aggression" priority over all other objectives. And yet, the Security Council in its first resolution regarding the conflict (resolution 479 dated 28 September 1980) called on Iran, a Member State of the United Nations, to practically submit to aggression. The

Secretary-General, on the other hand, has, through his subtle initiatives, endeavoured to gradually compensate for the Council's "purposeful error." The recent hypocritical political manoeuvre of the United States forced the Security Council to support the aggressor once again and hence openly violate the first and foremost objective of the Charter.

6. The Security Council, by virtue of its submission to the resolution presented and actively pursued by the United States, has, in practical terms, turned itself into a party to the conflict. As such, the Security Council will not be able to play a positive and constructive role as regards the war, and it will find it expedient in the future to modify this position.

7. The Islamic Republic of Iran had previously warned that the adoption of this resolution is a prelude to the expansion of tension and further exacerbation of the situation. What has transpired since its adoption points in the same direction. The United States immediately brought its armada into the Persian Gulf and intends to increase its military presence in the area. The incidents that took place in Saudi Arabia with American provocation and American military support for the export of Iraqi oil through Kuwait have soured the relations between the Islamic Republic of Iran and these countries. And military operations by Iraq are continuing unabated.

8. The United States increased presence and military provocations in the Persian Gulf which have led to further escalation of tension in the region constitute clear violations of paragraph 5 of resolution 598 (1987). As such, the United States is the first violator of the resolution whose formulation and adoption has been an American undertaking. If the Security Council is honestly committed to its own resolution, it follows that the Council must take a clearcut position on the violation of the resolution by United States.

9. The United States, through mounting pressure on the Security Council to adopt another resolution, seeks to take measures against the Islamic Republic of Iran and prepare the grounds for confrontation with Iran. Such efforts would only lead to the expansion of tension and further isolation of the Security Council from the political arena of the conflict. The Islamic Republic of Iran would not hesitate to confront any provocative measure on the part of the United States. However, the Security Council is well advised to realign its decisions with the principles enshrined in the Charter. Meanwhile, the best possible line of action lies in strengthening the positive aspects of the present resolution so that it would pave the way for future co-operation and adoption of constructive measures.

10. The crisis in the Persian Gulf is not unrelated to the question of the Iran-Iraq war. However, due to its significance and sensitivity, it calls for special attention. Failure in resolving the crisis in the Persian Gulf will precipitate the expansion of the conflict to unpredictable dimensions. Therefore, any efforts to restore peace and security in the Persian Gulf must be accorded first priority. In this connection, paragraphs 5 and 8 constitute the critical parts of the resolution.

11. The Islamic Republic of Iran and other countries in the region are, prior to and more than any other countries, interested in and committed to the stability and security in the Persian Gulf, freedom of navigation and the unimpeded flow of oil.

The crisis in the Persian Gulf was initiated, perpetuated and expanded by Iraq and brought to a climax following the Kuwaiti invitation of superpowers and consequent military intervention by the United States. All these developments, and in particular the violation of paragraph 5 of the resolution due to the intensified United States military presence in the area, have adversely affected the possibility of action on the part of the Secretary-General, on the basis of paragraph 8 of the resolution, in connection with security and stability in the region.

12. Resolution of the crisis in the Persian Gulf lies in the commitment of both parties not to attack commercial ships, withdrawal of foreign forces from the area and strict observance of neutrality on the part of all littoral States, particularly Kuwait. Only the restoration of tranquillity in the Persian Gulf holds out any hope for positive political action regarding other aspects of the conflict.

13. Iraq has violated the terms of resolution 598 (1987) by launching several offensive operations immediately following its adoption. Some instances of these violations have been reported to the United Nations by the Permanent Representative of the Islamic Republic of Iran and are contained in documents S/18997 and S/19002. The Iraqi aerial attack against Tabriz is another clear and unequivocal example of Iraqi violation of resolution 598 (1987) and its hypocritical policy, which, ironically, was engineered by Iraq's supporters, and to its best satisfaction. While expectation of a minimum of consistency between Iraq's words and deeds is most reasonable, it is not known for what reasons the Security Council has turned a blind eye in the face of such flagrant violations by Iraq of its resolution.

14. Pronouncement of Iraq as the aggressor and the party responsible for the conflict as well as determining damages and war reparations are essential for a thorough study of the conflict and formulation of a final solution. The Islamic Republic of Iran is ready to co-operate with the Secretary-General in this field.

15. Resolution 598 (1987) has condemned once again the use of chemical weapons, bombardment of civilian quarters, attacks on ships and civilian aircraft and violation of international law, particularly the Geneva Protocol of 1925 for the Prohibition of the Use in War of Asphyxiating, Poisonous and Other Gases. Yet, no practical measure has been foreseen to prevent the repetition of these crimes. The Islamic Republic of Iran is awaiting the response of the Secretary-General to the letter of the Minister for Foreign Affairs of the Islamic Republic of Iran dated 10 August 1987 (S/19029) regarding the important issue of use of chemical weapons particularly after the Iraqi chemical attack on the city of Sardasht. The Islamic Republic of Iran is also prepared to consider any proposals on other aspects of the war related to international humanitarian law.

16. The Islamic Republic of Iran has submitted to the Secretary-General of the International Committee of the Red Cross various proposals for repatriation of different groups of POWs and, in some instances, has acted on unilateral basis. The Islamic Republic of Iran is prepared to co-operate with the Secretary-General and the ICRC for implementation of these proposals or any other proposals on the basis of the Geneva Convention of 12 August 1949.

17. The eight-point plan of March 1985 of the Secretary-General has been the

only practical plan thus far taking various aspects of the war into account which has not been in force due to Iraqi opposition. This plan is still a suitable ground for future efforts of the Secretary-General.

18. The Islamic Republic of Iran hereby renews its confidence in the Secretary-General and is prepared to continue co-operation with him within the framework of his independent efforts and initiatives.

19. Constructive and commendable endeavours of certain impartial members of the Security Council to arrive at a balanced and positive resolution did not reach the desired results. However, grounds have been laid so that the Islamic Republic of Iran would continue its co-operation in a manner that would lead the Security Council to a just position. Undoubtedly, clearcut pronouncements on the responsibility of Iraq for the conflict, declared by some countries, constitute the most important element in the just resolution of the conflict. The Islamic Republic of Iran takes note of all positive endeavours and expresses its appreciation.

# APPENDIX VI

# The Ayat-Allah Khumayni's Address to the Islamic Mediating Committee (March 1, 1981)

*Imam Khumayni on March 1, 1981, met with the Delegation probing into the war imposed upon Iran by the Iraqi regime. At this meeting, after the statements made by Habib Chatti, the Secretary General of the Islamic Conference and spokesman of the delegation, Imam Khumayni delivered the following speech.*

## In the Name of God, the Merciful, the Compassionate

While thanking the gentlemen who have come here to observe Iran closely and what is happening here and to find out who is the oppressor and who has been oppressed and who is the aggressor and who has been aggressed against, if they have the time and intention to do so. If I intend to relate the events which have taken place and those which are going on in Iran and in this nation to you, even briefly, it will cause a loss in my health and also you do not have the time for this. Therefore, I will just speak to you about a number of issues.

I hope that Muslims, particularly the Muslim leaders, stop using the slogan of Islam as a cover for not following the tenets and precepts of Islam. They should think and practice Islam as it truly is. Up to now, the problem of Muslims and the Islamic oppressed nations has stemmed from this fact that their leaders have confined themselves just to Islamic slogans and under the cover of this, they have performed their other goals. I hope that the Islamic nations and especially their government, will transform these words into consciousness and into practicing Islam and the Holy Quran.

You have come to a country which has been under pressure and suppression for 2500 years; which has been trampled under the boots of dictatorial kinds, under the cover and pretext of justice, civilization and humanity and whose people were potentially seeking to free themselves of oppression and to acquire independence and freedom after being under pressure for many long years and having no freedom. Under the pretext of Islam and progress, their country had been trampled by the agents of the eastern and western powers. The selfishness and egoism of those who were affiliated to the superpowers did not give our people a chance to breath in freedom.

You have come to a country which has had about 60,000 martyrs and over 100,000 maimed and disabled as well as nearly 50,000 families who have lost their bread winners. You are now in a country which has witnessed all these miseries and has potentially about 1.5 million war refugees. They have been expelled from their cities and houses and have suffered various kinds of oppression. You have come to a country where the oppressors and tyrants have attacked due to their affiliation with the large powers. They (the enemy troops) suddenly crossed our borders and usurped a number of our cities by force. They killed all those young people who were present in those areas or took them prisoners.

You have come to a country in which our oppressed people have been subjected to attacks from every side such as military attack (the American military intervention in Iran's eastern part), the abortive coup plot and recently a military aggression by a puppet named Saddam Husein. All these attacks have been made merely because our tyrannized people intend to serve Islam, to return to Islam, to curtail the hands of the large powers which were active here in contrary to all human and Islamic criteria, to put an end to the rule of a dictatorial government and to establish the government it desires in this country and to live under the banner of Islam. The superpowers do not like to see the convergence of Islamic nations and governments. They are afraid of the unity of 1 billion Muslims under the banner of Islam.

You have come to a country in which the oppressed are confronting the oppressors, the aggressors and those who commit oppression and tyranny. You should not refer to the people of Iraq and Iran. These two people are brothers, are united and are against the Iraqi government. If you seek to count on Islam more than just talk, you have to visit and see what has happened to our cities, our youth, our women and our children. The remnants of our destroyed cities are now vivid proof of such crimes.

If you intend to transform your Islamic words into Islamic consciousness, if the Islamic governments seek to put words into practice and realization, they have to refrain from following selfish desires. They have to stop committing oppression and they should join the people.

If you stay here for a while and empty your ears from the propaganda against Iran and gather first-hand information about events taking place here, you will see the goals of the Iranian people and government. You will find out that what they want is Islam. They are seeking shelter in Islam. You will see that they do not want to propagate group interests or nationalism. They do not seek to evaluate the Iranians and devaluate the Arabs.

I should tell those individuals who attended the Taif Conference that they listened to what Saddam said for 80 minutes and not even for one minute did he speak to God's satisfaction. If he ever mentioned Islam, it was an Islam brought to him from Europe and America and not that Islam which is brought from Medina and the Hijaz. You listened to his nonsense for 80 minutes in which he claimed that Iran is the aggressor while his army was massacring the people inside our country; he still claims that Iran is an aggressor and none of you asked him to show where Iran has aggressed. Are we now fighting inside the Iraqi territory or are we struggling inside our own soil? If we are fighting inside our land, then the Iraqi (regime) is the aggressor and if one day we attack Iraq, then we will be the aggressor. When we defend the rights of our nation, we defend Islam and defend the rights of the people of Iraq and the Muslims. The Taif Conference should not sit idle and do nothing. You should not think that two nations are fighting with each other. The expectations of the Iraqi people are more than those of the Iranian people because the Iraqi nation has lost its religious leaders, its aged men, youths and children at the hands of this regime.

If you think of Islam, you should take the verses of the Quran into consideration. A verse says that if a group of Muslims, if they are truly Muslims, attacks another group of Muslims, all the Muslims are religiously bound to fight against the aggressor. If you practice this very verse, we do not expect anything more from you.

Probe into the aggression and if you do not have the time yourselves, appoint some representatives to go to the border areas which have been subjected to aggression. Send them to the graveyards which have been formed for us by them (the enemy troops). Dispatch them to visit the graves of our martyrs; the homeless families of the martyrs from the deprived and oppressed people of the south and the west of the country. They can go and probe into the case as to which side is the aggressor. If they find out that we began the aggression, then you can wage a war against us. And if you recognize that they (the Iraqi forces) aggressed, you wage a war against them (according to the Quran). Peace between Islam and atheism is meaningless. No Muslim should think that a peace must be created between Islam and the Muslims, on the one hand, and non-Muslims, on the other. The order of God should prevail and be followed. We all should follow the Holy Quran.

Do not think that today a dictatorial government is ruling Iran so that it can reach a peace accord with a person who the people are against. Do not imagine that here there is a dictator president who can talk with others against the will of the nation. Here the votes and belief of the nation rules. Here the nation itself has the rein of affairs in its hands. All the organs have been assigned by the people. Violating and neglecting the order of the people is not possible for any of us. It is impossible.

If you are assigned to put an end to this war, which is the ultimate goal of all Muslims, you should bring the aggressor to stand court trial and punish the invader. You should ask those who are in our country (have captured Iranian cities) to leave here and force Saddam to withdraw his forces from our country and his army to stop its invasion. After the aggression is stopped by him, then an international body can be held somewhere to prove into the crimes which have been committed. If we are criminals, we should be punished and if Saddam is, then he should be punished. This is the method of Islam. Islam has taught us correct orders. It has told us to be united together and refrain from any sort of division and discord in reality and not in words. We should not suffice ourselves to just holding gatherings and seminars and to say that a reality should be actualized.

Look at this conference which was held in Taif. What kind of measures were taken in regard to the oppression and tyranny being made upon us, in Palestine and Lebanon? What has been done for the sake of Muslims? We should not just assemble and cry out that we are Muslims and cry for Islam. The cries of Muhammad Reza were even more vocal. And now, Saddam as well as most of those who have their tyrannical and unjust rule over Islam and the Islamic states use the same slogans. These cries are being voiced by all. But if you intend to probe into the case and if you are truly with good will coming here to extinguish this fire, you should visit the borders; study the two sides, observe the two nations and then sit and judge. You should observe the Iraqi people to see whether they accept their government or not and also ask the same question from the Iranian people.

If the people approve their governments, then the latter are legitimate and if not, they are illegitimate. In that case, the world should take measures, if they (the world's people) are true in their claim that people should rule and that human rights should be respected. You have to refer to the people and we are prepared for you to come here to hold a referendum. Appoint observers from among yourselves for this referendum and ask the people whether or not they approve of their government. And you do the same in Iraq providing the bayonets are first removed and suppression is done away with. Hold a referendum there as well. Then you will see if the Iraqi people will say the same thing that the people of Iran say. (You will see) if they approve of it or not. If the Iranian people do not approve of its government, then you can tell the government to sit aside the people themselves take actions. If not, you go to Iraq and hold the same referendum there and ask the people whether they accept this party (the Ba'th party), this assembly of the party, this Saddam and this so-called president or not.

If the people approve of them, then we can sit and reach a compromise. But if the people do not approve, we do not accept either. Nobody accepts them but a group around him and the Soviet Union and the United States. If the issue is the same, then our gatherings will have no use and we will get nothing out of our talks. All of us should go under the banner of Islam, not in words, but in practice and in reality. If we verily go under the banner of Islam, then we will be able to act. Otherwise if there is just formalities such as having a formal meeting one day in Taif, one day in Iran and another day somewhere else, the issue will remain the same to the end and Muslims will be under the pressure of foreigners forever. Then we will remain defeated and oppressed to the end. We will rid ourselves from the oppression of the superpowers. With one billion people and vast recourses, when we find ourselves, we will find Islam and return to Islam and act according to God. If we follow the orders, God, the Almighty, will also help us. If you help God, then He will help you and will strengthen you.

Peace be upon you and the mercy and blessings of God.

# APPENDIX VII

# President Saddam Husayn's Fourth Message to the Iranian People (June 14, 1985)

**In the Name of God, the Compassionate, the Merciful,**

**Peace be upon those believing in peace,**

I have previously addressed to you messages based on the same intentions and purposes of this message, namely to express the desire of Iraq's people and leadership to achieve peace and good neighbourliness with Iran without any conditions which may encroach upon the sovereignty, security and dignity of Iran's peoples or armed forces.

Iraq's leadership as well as its great people and heroic armed forces have no designs on the dignity and sovereignty of Iranian peoples or their right to determine their own destiny. Since the outbreak of the war we have been incessantly calling for peace. Here we stress again that the peace we want is that on which relations between peoples and states in the world are based and which is respected and upheld by international law and organizations. It is the peace which is based on the principles of non-intervention in people's internal affairs and on respect for their sovereignty and choices. We challenge your rulers if they can cite any statement by any Iraqi official violating these principles since the outbreak of the war and even before that. However, you hear every day, just as Iraqi people do, statements by your rulers, with whom you are afflicted, which contradict these principles. It is your rulers who want to occupy Iraq, change its government and turn its free people into slaves. These expansionist intentions were the fatal beginning of the Iranian rulers' policy. Since the beginning these rulers, incited by Zionism and those fishing in troubled waters, fell into a trap they set for themselves when they have been driven by a primitive instinct and conceit about their power imagining that the victory of Iraq's peoples over the Shah could be turned into a means of aggression against the peoples of the region and even the world. Instead of thanking Almighty God for the victory of Iran's peoples over the Shah, carefully considering the factors leading to this victory and being modest in their relations with the world, their neighbours and the Iranian peoples, they have been so blinded by vanity that they imagined they could topple the governments in the region from outside and annex other countries to Iran. It was this aggressive beginning which led to the eruption of the war between Iran and Iraq five years ago which crushed thousands of people, dissipated numerous hopes and aspirations and brought to fruition many aims of the enemies of both the Arab and Islamic nations.

One of the basic qualities of a reasonable man is to opt for a new approach taking into account existing conditions, when he fails to deal with an issue because of using the wrong approach. It is not enough to examine the past period of the war. What is needed is to find a means to put an end to it. It is not enough to recall the losses and sacrifices but to put an end to those losses.

The task of leaders is essentially the application of principles. One of the requirements of this is that those who apply principles should take into account the existing conditions and resources. Those who fail to do so lose this quality, thus forfeiting a fundamental element of leadership.

The whole word including Iran's peoples, the Arab nation and Moslems all over the world have been telling Iran's rulers that the intentions and assessments which prompted them to start the war and insist on continuing it are illegitimate intentions and unrealistic assessments. Indeed, these assessments are impossible. Iran's rulers, however, are adamant in their arrogance risking the lives of Iran's peoples, and its economy and security as well as the opportunity to live peacefully.

I address you again proceeding from a high sense of responsibility and desire to secure peace for the peoples of Iran and Iraq. To understand fully the intentions and accuracy of this letter, you are requested to read as well my previous letters to you, to examine rationally all the principles, facts and forecasts contained in them. This would assure you that deception is neither in our code nor is it our means to achieve our objectives. Our basic code lies in the straightforward line of honest principles for the service of our people, world peace and legitimate aspirations of peoples.

Everything I said in my previous letters has been proved to be true and honest, especially with regard to explaining the intentions of your rulers in insisting on pursuing the war. We said the betting by Iran's rulers on prolonging the war was stupid and futile, that the war of attrition would be very costly for Iran and that they would eventually fail to achieve their wicked aims against Iraq. In this message I reiterate that, although the war has caused losses to the Iraqis and damage to the Iraqi economy, life in our country is prospering and progressing in all its aspects and fields. To continue to take up arms is an honourable mission, for Iraqis, because it is the only choice they have.

As for Iran, it has been paying a very high price as a result of the continuation of the war. It has been losing more of its citizens in this war and is losing its important opportunity in life. Furthermore, it has been suffering more destruction in economic and political terms. Its standing in the international public opinion as well as in the region has dropped to the lowest level. If putting an end to the war means stopping the losses in certain fields of life in Iraq, it obviously means saving life in Iran. Thus, true patriotism for Iranians would be opposition to the war, because by stopping the war Iranians would achieve all these objectives without risking their territory, dignity or security.

Iranian rulers' evil dreams to occupy Iraq and set up a surrogate treacherous regime, with God's aid, will only meet with failure. Iraq people will triumph, because God is with them and He does not like aggression and aggressors.

To assure you that the aim of Iraq's shelling deep inside Iranian territory is not intended to harm Iranian peoples physically nor is it intended to encroach upon their sovereignty or their dignity, but is rather a means of war to put an end to the war itself, and in order to provide suitable conditions during the days of the Lesser Bairman (Id Al-Fitr), we will stop shelling selected targets in towns deep inside Iranian territory for a specific period starting from 8 a.m. on Saturday morning 15 June 1985 to 30 June 1985.

Our decision to halt the shelling is intended to give your rulers another opportunity to consider peace and to give you another opportunity to press them to do so with a view to putting an end to the war and establishing peace. It comes in response to the calls of Iranian opposition forces which believe in peace between the two countries and which have asked us to stop shelling during this period to provide them with an opportunity to consolidate their campaign for peace.

In our view the practical bases to end the war are the following:

1. A ceasefire on land and sea and in the air.
2. Withdrawal of troops to international borders.
3. Total exchange of prisoners of war.
4. Conducting direct or indirect negotiations between Iraq and Iran to conclude an agreement of peace and good neighbourliness on the basis of respect for sovereignty and non-intervention in the internal affairs of either country. Direct negotiations may be preceded by preparatory contacts by a third neutral party.
5. The above principles are interrelated and the violation of any one of them constitutes violation of them all. They should also be implemented in accordance with practical steps and careful timing which does not allow for circumvention or evasion of commitments.
6. It is only natural to state that both Iraq and Iran should play a positive role in the region's security and stability so that peace would be final and not merely a provisional stage dictated by certain conditions.

    Do you find therefore, in any of the above principles—which we intend for stopping the war and putting a final end to it—any infringement of Iran's sovereignty, security and dignity?

It is obvious that these are fair principles ensuring sovereignty, security and dignity for the people of Iraq as well as for the Iranian peoples. They are derived from the international law, the United Nations Charter and the people's right to life, independence and freedom.

This is what we propose to you and on this basis we will stop shelling operations as we have just said. This may be another opportunity for you to exercise strong pressure on your rulers to accept ending the war and resorting to peace. It may also provide an appropriate opportunity for your rulers to use whatever reason is left in them to accept peace.

However, it is necessary to point out that the practical conditions of the war which your rulers insist on imposing on us and on you demand that we should set certain rules for stopping the shelling. We will be forced to resume shelling before the end of the period set above in the following cases:

1. If the Teheran rulers continue to shell our towns and villages.
2. If your rulers launch an offensive on our territory.
3. If they issue statements rejecting our initiative and calling for war and killing as well as threats.
4. If they deploy enemy forces inside towns for the purpose of launching offensives on our territory and armed forces.

In such circumstances we will be forced to resume shelling and striking hard before the end of the period set to teach those who advocate war and destruction the lessons they deserve. it is also to show everybody that occupation of Iraq and imposing humiliating conditions on it with force is sheer imagination and that the only path open is that of peace.

From the position of strength, I can confirm to you our true desire to live in peace based on just and honourable bases as well as those of good neighbourliness. I also stress that continuation of the war by your rulers will bring about nothing but destruction and ruin for you, and disappointment and defeat for your rulers. We hope that you will succeed in pressing these rulers to give up their deviating and evil policies and in finding a new path for Iran and the peoples of the region, namely that of peace, security and stability. If your rulers insist on continuing the war, killing and intimidation, and if they do not respond positively to our initiative, we will be forced to continue to strike by all possible means including the shelling of selected targets wherever they may be. This will be in defence of our people and sovereignty. He who warns is forgiven.

> Saddam Husayn
> President of the Republic of Iraq
> Baghdad, 25 Ramadhan 1405 A.H.
> 14 June 1985

# Index

231